APPAREL QUALITY
Lab Manual

APPAREL QUALITY
Lab Manual

JANACE E. BUBONIA, PH.D.
TEXAS CHRISTIAN UNIVERSITY

FAIRCHILD BOOKS
AN IMPRINT OF BLOOMSBURY PUBLISHING INC
B L O O M S B U R Y
NEW YORK • LONDON • NEW DELHI • SYDNEY

Fairchild Books
An imprint of Bloomsbury Publishing Inc.

1385 Broadway
New York
NY 10018
USA

50 Bedford Square
London
WC1B 3DP
UK

www.bloomsbury.com

FAIRCHILD BOOKS, BLOOMSBURY and the Diana logo are trademarks of Bloomsbury Publishing Plc

ISBN: 978-1-62892-4572

Typeset by Precision Graphics
Cover Design by Eleanor Rose
Printed and bound in the United States of America

TABLE OF CONTENTS

INTRODUCTION TO THE LAB MANUAL

The purpose of the *Apparel Quality Lab Manual* is to reinforce the chapter contents and lecture material from *Apparel Quality: A Guide to Evaluating Sewn Products* through the integration of hands-on lab activities and projects to enhance learning and engage student's critical thinking skills. This lab manual is presented in a user-friendly format that is easy to read and understand. Lab activities and projects allow for greater understanding and better recognition of quality issues that arise with apparel production and end use.

Today's student wants to be actively engaged in their learning, and the projects and lab activities allow them to actively participate in determining the quality level of apparel products. The projects and activities also allow students to simulate methods used for analyzing quality in today's fashion industry. This lab manual includes activities for analyzing products, testing and evaluating materials and garments, project sheets for product comparison testing, worksheets to record data, directions for mounting specimens after testing, templates for cutting specimens, and inspection forms for evaluating product defects.

The lab assignments reinforce information learned in the chapters of the main text. It is important to note that not all programs have equipment for textile testing; therefore, the manual includes both product analysis labs as well as testing activities to enhance learning to meet the needs of both types of classroom and laboratory environments. Some programs have testing labs complete with testing equipment, whereas others do not. Therefore, based on the availability of equipment, instructors can determine which project components and lab activities their students will complete. The lab activities and projects are designed to meet the needs of courses focused on visual inspection as well as those that conduct physical testing. This manual can also be utilized as a standalone resource to supplement a course with hands-on relevant activities.

When used in conjunction with the main text *Apparel Quality: A Guide to Evaluating Sewn Products*, there are lab components for every chapter except Chapter 8 Sourcing, Assembly, and Mass Production of Sewn Products. The activities at the end of this chapter adequately reinforce the concepts presented in the text. The *Apparel Quality Lab Manual* identifies the main text chapter(s) that corresponds with the lab manual at the beginning of every chapter.

The piece goods and Comparison Project lab activities span several weeks so they will integrate information learned from many different chapters. These projects are intended to simulate real-world garment analysis and material-testing scenarios. Based on the availability of equipment, instructors can determine which project components and lab assignments their students will complete. Through these analyses, students will gain a better understanding of the causes for product failure and customer dissatisfaction. They will also learn how product development and manufacturing decisions influence the quality of end products.

CORRELATION BETWEEN BOOK AND LAB MANUAL CHAPTERS AND ACTIVITIES

The following table shows the *Apparel Quality Lab Manual* chapters and activities and their corresponding chapters from this text.

TEXT CHAPTER	LAB MANUAL CHAPTER	LAB ACTIVITIES
1	1	1.1 Customer Expectation Survey for Jeans 1.2 Quality Cues Comparison 1.3 Product Selection 1.4 Comparison Project Customer Expectation Survey
2	2	2.1 Analysis of Aesthetic and Design Details of Woven Garments in Relation to Design Elements and Principles across Price Categories 2.2 Comparison Project Product Analysis of Aesthetic and Design Details in Relation to Design Elements and Principles of Competing Brands 2.3 Farnsworth Munsell 100 Hue Test
3, 11	3	3.1 Raw Materials Classification 3.2 Specimen Templates and Sampling Plan 3.3 Characterization Testing on Piece Goods 3.4 Characterization Testing on Comparison Project Garment 3.5 Wear Testing Comparison Project Garments
4	4	4.1 Garment Analysis of Construction Details across Price Classifications 4.2 Comparison Project Garment Analysis of Construction Details across Competing Brands 4.3 Comparison Project Garment Evaluation of Appearance and Color Change after Laundering
5	5	5.1 Comparison Project Garment Analysis of Size and Fit
6, 7	6	6.1 Comparison Project Garment Identification of Stitches and Seams 6.2 Comparison Project Garment Thread Consumption Calculations
9	7	7.1 Comparison Project Garment Label Compliance 7.2 Garment Care
10	8	8.1 Garment Safety and Compliance
11, 12	9	9.1 Test Methods for Evaluating Selected Garments 9.2 Piece Goods Appearance and Performance Testing 9.3 Comparison Garment Appearance and Performance Testing 9.4 Comparison Project Garment Results and Performance Specifications
13	10	10.1 Inspection of Randomly Selected Garments 10.2 Comparison Project Garment Inspection 10.3 Comparison Garment Customer Satisfaction Survey

CHAPTER 1

Apparel Quality and Consumer Perceptions Lab

LAB OBJECTIVES:

- To gain a better understanding of quality, its basic elements, and how it is measured through hands-on activities.
- To examine factors that impact consumer perceptions and expectations of apparel products before and after purchase.
- To identify quality cues pertaining to a selected apparel product.

In Chapter 1 of *Apparel Quality: A Guide to Evaluating Sewn Products*, the basic elements of quality are introduced, as are the factors impacting consumer perceptions and expectations of apparel products. It is important to understand what quality is and how it is evaluated. Furthermore, one must consider the relationship between customer expectations and perception of quality before and after purchase. These factors are very important to brands when developing, producing, and distributing products.

There are four lab activities (worksheets 1.1, 1.2, 1.3, 1.4) for Chapter 1 as well as guidelines for the comparison project. They will focus on activities relating to customer expectations and perceptions of quality. Selection of the comparison project garment will be decided during this lab session and further analysis and testing will be completed over several lab sessions. Lab Activities 1.1 and 1.2 should be completed during this lab session. Comparison project garments for Lab Activity 1.3 must be purchased and brought to the next lab session. Lab Activity 1.4: Comparison Project Customer Expectation Survey should also be completed during the first lab session.

PREPARATION AND SUPPLIES FOR CHAPTER 1 LAB ACTIVITIES

Students will form teams of two to three people for small group work and teams of four to five people for the Comparison Project and piece goods testing. These teams will remain the same throughout the course of this class. Supplies are needed for lab activities, and there may also be things that need to be completed or prepared for the next lab session. The following discussion of each chapter will provide information regarding the supplies and garments that are needed for each of the lab activities. Some labs will require preparation or completion of wear-testing and refurbishment in order to complete chapter activities. Other lab activities may require groups to meet prior to a lab session to make decisions regarding garments that each member will bring to class for evaluation.

In preparation for the Chapter 1 lab activities, meet with your small group (two to three people). Each group member will bring one pair of jeans from his or her wardrobe to lab for evaluation. A minimum of two price categories (budget, moderate, better, contemporary, bridge, designer) should be represented within each two-person lab group, and a minimum of three price classification are needed for each three-person lab group. Determine which price classification will be represented by the jeans each person will bring.

Lab Activities 1.1 and 1.2: Quality Analysis of Jeans across Price Classifications

For Lab Activities 1.1 and 1.2, you will work in teams of two to three people to analyze the quality of jeans across price classifications. Each group member will bring one pair of jeans from his or her wardrobe to lab for evaluation. A minimum of two price categories should be represented within each two-person lab group or a minimum of three price classifications for each three-person lab group (budget, moderate, better, contemporary, bridge, designer). As a team, begin completing the Customer Expectation Survey for Jeans worksheet 1.1. Discuss and record your group's expectations, preferences, and quality perceptions for each pair of jeans and the corresponding price classifications. Once the Customer Expectation Survey for Jeans is completed move on to Quality Cues Comparison worksheet 1.2.

Each team member will complete the Quality Cues Comparison worksheet 1.2 by visually inspecting and evaluating his or her own pair of jeans. Information pertaining to intrinsic and extrinsic quality cues and to the elements of quality that use the user-defined approach will be recorded. See Table 1.1: Quality Approaches, Elements, Characteristics, and Cues to assist you with completing the worksheet for Lab Activity 1.2. As you complete this worksheet, keep in mind the ratings assigned for your price classification in the Customer Expectation Survey for Jeans (Lab Activity 1.1).

Lab Activities 1.3 and 1.4: Comparison Project Overview

In the industry, companies will purchase competitors' products and evaluate them in comparison to their own. This investment of resources is made in an effort to determine how their brands' garments can be improved and how they stack up against the competition. The Comparison Project will simulate this process. It is recommended for lab groups to be comprised of four to five students.

Supplies and Product Selection

Each lab group will test t-shirts, tank tops, or Henleys. Discuss which type of garment you would like to test for the Comparison Project. After the group has determined

Table 1.1 Quality Approaches, Elements, Characteristics, and Cues

APPROACHES TO QUALITY	
Product-Defined	Focuses on measureable physical features and attributes. Inherent physical features help determine the overall quality level.
Manufacturer-Defined	Focuses on meeting specifications for conformance to production standards for customer satisfaction.
User-Defined	Focuses on meeting customer needs for aesthetics, performance, and function in relation to purchase price and value.
ELEMENTS OF QUALITY	
Performance	Garment's functional aspects and features for its intended use. Garment Features are directly linked to performance: Fibers Yarns Material structure Seam construction Fabric or garment finishes Product-defined and user-defined quality approaches
Durability	Garment's ability to resist physical and mechanical deterioration and function for the useful life of the product. Product-defined and user-defined quality approaches
Serviceability	Garment's ease of care, ability to retain its shape and appearance, and cost of care and repair User-defined quality approach
Conformance	Garment's degree to which it meets standards for design, materials, and product specifications Design Specifications relate to aesthetic appeal: Styling details Design features Style characteristics Manufacturer-defined quality approach
Aesthetics	Garment's ability to engage the senses and meet consumer personal preferences Aspects of Aesthetics: Appearance Comfort Sound Smell User-defined quality approach

continued

Table 1.1 *(cont.)*

QUALITY CHARACTERISTICS AND CUES	
Primary Quality Indicators/Physical Attributes, Performance Features, And Intrinsic Quality Cues	Product benefits Tangible attributes assessed by Sight Touch Sound Smell Physical Attributes: Overall attractiveness of materials Garment style and construction details Appropriateness of garment design for end use Performance Features: Fit Durability Effectiveness for use Ease of care
Secondary Quality Indicators/Extrinsic Quality Cues	External factors that influence consumers' perception of quality—not part of the physical makeup of the product External cues: Retail price Brand reputation Advertising and marketing Visual presentation in store/online/catalogue

the type of garment, select the style (i.e., short sleeve, long sleeve, v-neck, scoop neck, spaghetti strap, racer back). Now determine the color, type of knit fabric (i.e., jersey, rib, interlock), and fiber content of the garment. All members will test different brands of the same style garment in the same color, same fabric, and same fiber content. For example, a group may decide to test purple, 100 percent cotton jersey knit, short sleeve, scoop neck t-shirts from the following brands: Apt. 9, Gap, Hanes, Pink, and Mossimo.

Each team will make a list of competing brands (e.g., Apt. 9, Canyon River Blues, Faded Glory, Gap, Hanes, jcp, Merona, Old Navy, Route 66, Simply Vera, SONOMA life, Xhilaration) and divide them among its members. Each person should have two brands to choose from. This will help ensure that individuals within a team are not purchasing the same brands. Do not purchase garments from off-price retailers because quality issues could be the reason they are sold there. Each lab group member is responsible for purchasing three identical retail products to be evaluated and wear-tested (one shirt to wear test, one shirt to cut specimens for conducting physical testing and deconstruct for analysis of stitches and seams, and one shirt to act as a control). It is imperative for each individual to purchase three identical garments (exact same color, size (must be designated for your sex-male or female and fit true to size), fabrication, fiber content, country of origin, finish, brand, price). Complete Lab Activity 1.3 Product Selection worksheet. ***Garments must be purchased and brought to the next lab session (Lab 2).***

Guidelines

Data will be recorded on comparison worksheets 1.3, 1.4, 2.2, 3.4, 3.5, 4.2, 4.3, 5.1, 6.1, 7.1, 9.3, 9.4, 10.2, and 10.3. The availability of testing apparatuses and equipment will determine which of the following tests will be completed for the comparison project: yarn construction, fiber identification or classification, fabric construction, count, weight, dimensional change, garment twist/skewness, appearance, color permanence, tensile strength-bursting, abrasion-pilling, cost per wear, comfort and fit, quality of construction, stitching, seams, defects, overall performance, and value.

Each lab group member is responsible for wear-testing the retail product he or she purchased. Before any testing begins, each student is required to have a team member take a "before" photo of the garment. Photos must be good quality. Please note location of photo, distance from camera, and light source. The "after" photo must be taken in the same place, distance, and with the same lighting condition. Wear-testing will require each student to do the following:

- Wear the garment once a week, for a minimum of four hours, maximum of six hours per day or night for five weeks (depending upon garment selected and intended use—only sleepwear items should be slept in).
- After the garment has been worn, it must be refurbished according to the procedures outlined on the care label.
- The garment must be worn five times and refurbished five times and brought to your weekly designated lab session to perform testing and evaluation.

The Comparison Project Garment Results and Performance Specification worksheet 9.4 will be completed by each team member. Lab test data will be recorded and compared to industry standard specifications for evaluating the selected product. Any images contained within the Comparison Project worksheets must be printed in color—all other information is acceptable in black and white.

Once your team has determined the garment style you will evaluate, collaboratively complete the Lab Activity 1.4: Comparison Project Customer Expectation Survey worksheet. Each team member should record the data on his or her own worksheet because this will be turned in with the project forms for your garment brand.

Lab Activity 1.1: CUSTOMER EXPECTATION SURVEY FOR JEANS

Each team member, in collaboration with the rest of the team, will complete this worksheet for his or her own jeans.

Your Name

Name(s) of Teammate(s)

Product Description	**Price Classification**

Rank the importance of each item on a scale of 1–5.
5 = Extremely Important *4 = Very Important* *3 = Moderately Important*
2 = Slightly Important *1 = Not Important*

Primary Quality Indicators/Intrinsic Quality Cues

Aesthetic Characteristics	**5**	**4**	**3**	**2**	**1**
Overall attractiveness of materials, styling, and design of garment for its intended use					
Maintains shape and appearance during wear					
Maintains shape and appearance after refurbishment (cleaning)					
Functional Characteristics	**5**	**4**	**3**	**2**	**1**
Ability of the garment to perform for its intended use					
Comfort of the garment					
Fit of the garment					
Accuracy of sizing/True to size					
Durable					
Color maintained after cleaning; color does not rub off or bleed onto other items					
Ease of care					

Lab Activity 1.1 *(cont.)*

Secondary Quality Indicators/Extrinsic Quality Cues

External Factors	5	4	3	2	1
Price					
Value					
Brand					
Country of origin					
Marketing, advertising and media presence					
Visual presentation					

Answer the following questions about your team's preferences for jeans.

What fiber content do you prefer for jeans?

How often do you wear this price classification of jeans?

How long do you expect a pair of jeans within this price classification to last (useful life of the product)?

What aesthetic expectations do you have for jeans within this price classification?

Lab Activity 1.1 *(cont.)*

What performance expectations do you have for jeans within this price classification?

What expectations do you have regarding price and brand for jeans within this price classification?

Lab Activity 1.2: QUALITY CUES COMPARISON

Each team member, in collaboration with the rest of the team, will complete this worksheet for his or her own jeans.

Your Name

Name(s) of Teammate(s)

Product Description	Brand
Price Classification	**Retail Purchase Price**

Briefly describe each of the physical attributes and performance features based on your perception of the jeans.

Primary Quality Indicators/Intrinsic Quality Cues

Physical Attributes	
Fiber content	
Fabric weave (plain or twill)	
Appropriateness of fabric weight for garment and intended use	
Color and wash description	
Findings and trim components	
Thread and topstitching	
Styling and design of garment for its intended use	
Garment construction	
Maintains shape and appearance during wear	
Maintains shape and appearance after refurbishment (cleaning)	

Lab Activity 1.2 *(cont.)*

Performance Features	
Ability to perform for intended use	
Comfort of the garment	
Fit of the garment	
Accuracy of sizing/true to size	
Durability for intended use	
Ease of care	

Briefly describe each of the external factors and how they influence your perception of the quality of the jeans.

Secondary Quality Indicators/Extrinsic Quality Cues

External Factors	
Price	
Brand reputation	
Country of origin	

Calculate the cost per wear for your jeans.
Purchase price ÷ Number of times worn = Cost per wear

Cost Per Wear Calculation

Cost Per Wear	
Purchase price	
Number of times worn	
Cost per wear	
How long have you owned the jeans?	

Lab Activity 1.2 *(cont.)*

Briefly describe each of the elements of quality (see Table 1.1) as they relate to the jeans and if each meets your needs. Explain why or why not. Rate the level to which your needs are met by the jeans on a scale of 1–3.

3 = Completely meet *2 = Sort of meet* *1 = Do not meet*

Elements of Quality That Use the User-Defined Approach to Quality

Performance	Rating and Explanation of Why Needs Are Met or Not Met
Durability	**Rating and Explanation of Why Needs Are Met or Not Met**

Lab Activity 1.2 *(cont.)*

Serviceability	Rating and Explanation of Why Needs Are Met or Not Met

Aesthetics	Rating and Explanation of Why Needs Are Met or Not Met

Lab Activity 1.2 *(cont.)*

Now that each team member has evaluated his or her garment, discuss your evaluations and determine which pair of jeans is the best quality garment? Which garment is the best value?

Best Quality Garment Brand	Best Value Garment Brand

Is the brand selected for Best Quality the same as the one chosen for Best Value? ☐ Yes ☐ No Explain your answer and provide justification for why you chose the brand(s)?

Lab Activity 1.3: PRODUCT SELECTION

Each team will document the garment style and details for the Comparison Project in this worksheet. Bring this with you when shopping for your garments.

Your Name

Name(s) of Teammate(s)

Product Selected	
Type of garment selected	
Garment style details	
Color	
Type of knit fabric	
Fiber Content	

Make a list of eight to ten brands (not stores). Each person will select two brands from the list. Indicate your first and second choice brands. No brand can be duplicated within a team.

Brands	**Brands**
Your First-Choice Brand	**Your Second-Choice Brand**

Lab Activity 1.4: COMPARISON PROJECT CUSTOMER EXPECTATION SURVEY

Each team will collaboratively complete this survey for the selected garment style. Each individual team member will record the information on his or her own worksheet.

Your Name

Name(s) of Teammate(s)

Product Description	**Price Classification**

Rank the importance of each item on a scale of 1–5.
5 = Extremely Important *4 = Very Important* *3 = Moderately Important*
2 = Slightly Important *1 = Not Important*

Primary Quality Indicators/Intrinsic Quality Cues

Aesthetic Characteristics	**5**	**4**	**3**	**2**	**1**
Overall attractiveness of materials, styling, and design of garment for its intended use					
Maintains shape and appearance during wear					
Maintains shape and appearance after refurbishment (cleaning)					
Functional Characteristics	**5**	**4**	**3**	**2**	**1**
Ability of the garment to perform its intended use					
Comfort of the garment					
Fit of the garment					
Accuracy of sizing/true to size					
Durable					
Color maintained after cleaning; color does not rub off or bleed onto other items					
Ease of care					

Lab Activity 1.4 *(cont.)*

Secondary Quality Indicators/Extrinsic Quality Cues

External Factors	5	4	3	2	1
Price					
Value					
Brand					
Country of origin					
Marketing, advertising and media presence					
Visual presentation					

Answer the following questions about your team's preferences for the garment.

What fiber content do you prefer for this type of garment?

How often do you wear this type of garment?

Lab Activity 1.4 *(cont.)*

How long do you expect this type of garment to last (useful life of the product)?

What aesthetic expectations do you have for this type of garment?

What performance expectations do you have for this type of garment?

What expectations do you have regarding price and brand for this type of garment?

NOTES

CHAPTER 2

Integrating Quality into the Development of Apparel Products Lab

LAB OBJECTIVES:

- To identify design details of garments that comprise the overall aesthetic appeal.
- To analyze garments and identify the design elements and principles as they are applied to selected apparel products.
- To determine your color visual acuity.

In Chapter 2 of *Apparel Quality: A Guide to Evaluating Sewn Products*, the process used for designing and developing an apparel line is presented in the logical order in which it occurs in the fashion industry. An overview of the design elements and principles was presented along with discussion of how they are used to create aesthetically pleasing salable products consumers want to buy. Color management and methods for effectively communicating color throughout the design and manufacturing stages is important. It is also critical for individuals evaluating color to possess high visual acuity so hues are accurately evaluated.

There are three lab activities (worksheets 2.1, 2.2, 2.3) for Chapter 2. Labs will focus on analyzing garments and identifying aesthetic and design details and the related design elements and principles applied to selected garments. The lab activities in this chapter should be completed during one lab session or prior to the next lab in the case of worksheet 2.3. For Lab Activities 2.1 and 2.2, you will work in Comparison Project teams of four to five students. For Lab Activity 2.3, you will work independently.

PREPARATION AND SUPPLIES FOR CHAPTER 2 LAB ACTIVITIES

In preparation for the Chapter 2 lab activities, you will meet with your small group (two to three people). Each group member will bring one woven shirt from his or her wardrobe to lab for evaluation. A minimum of two price categories (budget, moderate, better, contemporary, bridge, designer) should be represented within each two-person lab group, and a minimum of three price classification are needed for each three-person lab group. Each person will also bring all three of his or her comparison garments to lab.

Lab Activity 2.1: Analysis of Aesthetic and Design Details of Woven Garments in Relation to Design Elements and Principles across Price Categories

For Lab Activity 2.1, work in teams of two to three people, and each group member will bring one woven shirt from his or her wardrobe to lab for evaluation. A minimum of two price categories should be represented within each two-person lab group or a minimum of three price classifications for each three-person lab group (budget, moderate, better, contemporary, bridge, designer). As a team, begin completing the worksheet for Lab Activity 2.1: Analysis of Aesthetic and Design Details of Woven Shirts in Relation to Design Elements and Principles across Price Categories. Discuss and record the aesthetic and design details, and evaluate the design principles and elements used. Use Table 2.1 and Figure 2.1 to help you complete the worksheet. After worksheet 2.1 is completed, move on to Lab Activity 2.2: Comparison Project Product Analysis of Aesthetic and Design Details in Relation to Design Elements and Principles of Competing Brands.

Table 2.1 Textile Finishes That Alter the Appearance and Texture of Fabrics

FINISH NAME	PROCESS	EFFECT
Calendering	Mechanical	Embossing Glazed appearance (low-high sheen) Moiré effect
Flocking	Mechanical	Patterned surface Suede or velvety texture
Fulling	Mechanical	Felting Improves hand, softness, and thickness
Mercerizing	Chemical	Luster Improves drape, hand, and strength
Napping and sueding	Mechanical	Brushed surface Improves hand, insulation, and softness
Plissé	Chemical	Puckered surface
Shearing	Mechanical	Trimmed surface fibers Consistency and uniformity
Softening	Chemical and mechanical	Improves drape and hand
Stiffening	Chemical	Improves crispness and stiffness
Washing	Chemical and mechanical	Distressed, faded, worn appearance Improves softness

Figure 2.1 Caption: Garment Silhouettes

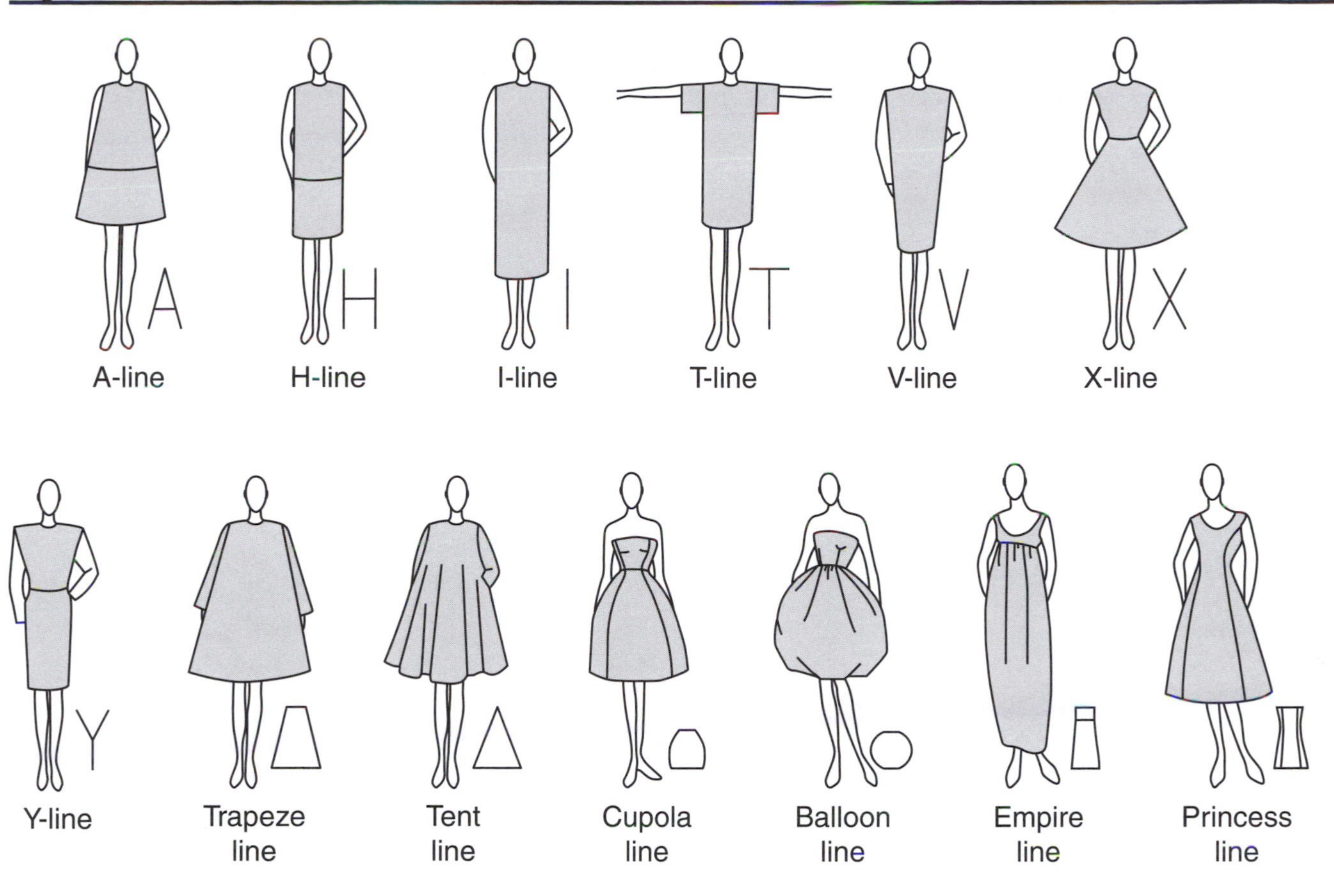

Lab Activity 2.2: Comparison Project Product Analysis

For Lab Activity 2.2, each group member will bring his or her Comparison Project knitted garments they purchased to lab for evaluation. This lab activity will be conducted in the Comparison Project lab groups (teams of four to five students). As a team, begin completing Lab Activity 2.2: Comparison Project Product Analysis of Aesthetic and Design Details in Relation to Design Elements and Principles of Competing Brands. Discuss and record the aesthetic and design details, and evaluate the design principles and elements used. Refer to Table 2.1 and Figure 2.1 to assist in completing certain portions of the worksheet.

Lab Activity 2.3: Farnsworth Munsell 100 Hue Test

You may need to complete this activity outside of class because it requires the use of a computer. Log onto http://www.colormunki.com/game/huetest_kiosk, and complete the Farnsworth Munsell 100 Hue Test to determine your visual acuity. This measures how well individuals can visually discriminate between fine distinctions of hues, and your score determines if you are able to see color well enough to rate in lab. A score of zero means you have perfect visual color acuity. The lower the score, the better your visual acuity is. You must score 28 or below to rate color during lab activities. If your hue test score is above 28, you can still participate in color-related lab activities; however, a teammate that can rate must assist with rating to ensure accuracy.

Lab Activity 2.1: ANALYSIS OF AESTHETIC AND DESIGN DETAILS OF WOVEN GARMENTS IN RELATION TO DESIGN ELEMENTS AND PRINCIPLES ACROSS PRICE CATEGORIES

Each team member, in collaboration with the rest of the team, will complete this worksheet for his or her own woven shirt. Document how each of the aspects on this worksheet are used to create the overall aesthetic appeal.

Your Name

Name(s) of Teammate(s)

Product Description—describe the style of the garment	Price Classification

Garment Aesthetic and Design Details

Color	
What color(s) is the garment?	
Do all areas of the garment match? If no, explain.	
Have you experienced metamerism with this garment? If so, explain how the color changes.	
Does thread match the fabric color? Is it supposed to or is it a contrasting color?	

Lab Activity 2.1 *(cont.)*

Texture and Pattern	
Describe the surface texture of the shirt?	
What type of fabric is the shirt made from?	
Is there an aesthetic finish that alters the physical texture of the shirt (see Table 2.1)?	
Does the shirt have a pattern? If so how is it achieved (i.e., weaving, printing)?	
Proportion, Balance, and Emphasis	
What is the silhouette of the shirt (see Figure 2.1)?	
Are the pieces that comprise the shirt proportionate and aesthetically pleasing (i.e., collar/neckline, cuffs, sleeves, length of garment, yoke)?	
Does the shirt have symmetric or asymmetric balance? Is the garment balanced? If not, describe how the garment design could be improved.	
What is the focal point of the shirt?	
Are all of the shirt portions and components harmonious? If not, explain.	

Lab Activity 2.1 *(cont.)*

Line	
In what ways has the designer used line to create the silhouette and shapes formed within the shirt (i.e., darts, seams, tucks, pleats, gathering, stitching, linear trims, piecing)?	
Does the viewer's eye flow from one part of the garment to the next with rhythm? This is controlled through the lines created by seams, darts, and other design details. If not, how can the design be improved?	

Answer the following questions about your team's findings after evaluating all of the shirts across price categories.

Are there any design details you would have expected to see in a shirt from a particular price category that was not present? Explain your answer.

Lab Activity 2.1 *(cont.)*

Were there any design details you were surprised to see in the shirts from a particular price category? Explain your answer.

Do the design details for each price classification correspond to your group's expectations for each price category? Explain your answer.

Lab Activity 2.2: COMPARISON PROJECT PRODUCT ANALYSIS OF AESTHETIC AND DESIGN DETAILS IN RELATION TO DESIGN ELEMENTS AND PRINCIPLES OF COMPETING BRANDS

Each team member, in collaboration with the rest of the team, will complete this worksheet for his or her own comparison garment. Document how each of the aspects on this worksheet are used to create the overall aesthetic appeal.

Your Name

Name(s) of Teammate(s)

Product Description—describe the style of the garment	**Brand and Price Classification**

Garment Aesthetic and Design Details

Color	
What color(s) is the garment?	
Do all areas of the garment match? If no, explain.	
Have you experienced metamerism with this garment? If so, explain how the color changes.	
Does thread match the fabric color? Is it supposed to or is it a contrasting color?	

Lab Activity 2.2 *(cont.)*

Texture and Pattern	
Describe the surface texture of the garment?	
What type of fabric is the garment made from?	
Is there an aesthetic finish that alters the physical texture of the garment (see Table 2.1)?	
Does the garment have a pattern? If so how is it achieved (i.e. knitting, printing)?	
Proportion, Balance and Emphasis	
What is the silhouette of the garment (see Figure 2.1)?	
Are the pieces that comprise the garment proportionate and aesthetically pleasing (i.e. neckline, trim, sleeves, length of garment, pockets)?	
Does the garment have symmetric or asymmetric balance? Is the garment balanced? If not, describe how the garment design could be improved.	
What is the focal point of the garment?	
Are all of the garment portions and components harmonious? If not, explain.	

Lab Activity 2.2 *(cont.)*

Line	
In what ways has the designer used line to create the silhouette and shapes formed within the garment (i.e. darts, seams, tucks, pleats, gathering, stitching, linear trims, piecing)?	
Does the viewer's eye flow from one part of the garment to the next with rhythm? This is controlled through the lines created by seams, darts, and other design details. If not, how can the design be improved?	

Answer the following questions about your team's findings relating to the competing garments.

Are there any design details you would have expected to see in the garments that were not present? Explain your answer.

Lab Activity 2.2 *(cont.)*

Were there any design details you were surprised to see in the shirts from a particular price category? Explain your answer.

Are the design details consistent across the competing brands or were there distinct differences? Explain your answer.

Lab Activity 2.2 *(cont.)*

Is there one brand that your group believes has a higher aesthetic appeal based on your findings? Explain your answer.

Lab Activity 2.3: FARNSWORTH MUNSELL 100 HUE TEST

Log onto http://www.colormunki.com/game/huetest_kiosk, and complete the Farnsworth Munsell 100 Hue Test to test your visual acuity.

Your Name

Time test started
Time test ended
Your score
Best score for your gender and age
Highest score for your gender and age
Is your score 28 or below?
Does your score allow you to accurately rate color?

NOTES

CHAPTER 3

Raw Materials and Sewn Products Testing Lab

LAB OBJECTIVES:

- To identify fiber types, yarn construction, fabric structures, and finishes.
- To understand fabric characteristics in relation to performance and selection for apparel products.
- To complete specimen templates and a sample plan for testing.
- To execute fabric characterization tests to determine fiber or fabric properties.
- To prepare garments for the Comparison Project garment wear test.

Chapters 3 and 11 of *Apparel Quality: A Guide to Evaluating Sewn Products* review fiber types, yarn construction, fabric structures, colorants, and finishes to provide an understanding of fabric characteristics as they relate to product performance. Raw materials selection and the factors that influence the quality of apparel items are important to know when developing, manufacturing, and selling merchandise. Fabric properties and characteristics influence designers' selection of materials for use in apparel items and can impact the customer's purchase and satisfaction level with the finished product. The basic components of test methods and fabric characterization testing are introduced in Chapter 11 to provide insight into how compliance with specification data is verified. Review Chapters 11 and 12 of *Apparel Quality: A Guide to Evaluating Sewn Products* because these concepts will be introduced in preliminary work in this lab and subsequent labs.

There are five lab activities (worksheets 3.1 through 3.5) for Chapter 3. You may complete all or only portions of the lab activities for this chapter—those for which your school has apparatuses or equipment. Lab activities will focus on raw materials review, characterization, and preliminary preparation for wear-testing your comparison product.

PREPARATION AND SUPPLIES FOR CHAPTER 3 LAB ACTIVITIES

In preparation for Chapter 3 lab activities, each team of four to five students will purchase 2 yards or 2 meters of woven fabric of a medium to dark color that will be used for piece goods testing. In addition, groups will need four pages of card stock or one package of Visi-GRID Quilter's Template Sheets (contains four plastic grid sheets,

8 ½ inches by 11 inches or metric equivalent), graph paper, a permanent marking pen (such as Sharpie rub-a-dub), a ruler, scissors, and two additional pieces of card stock for mounting specimens. Each person will also bring all three of his or her comparison garments to lab for preparation and testing.

Lab Activity 3.1: Raw Materials Classification

Lab Activity 3.1: Raw Materials Classification will focus on review of raw materials composition. This lab activity can be completed during this lab session or prior to the next lab session. For Lab Activity 3.1, work independently or in teams of two to three people. Complete the raw materials worksheet 3.1. Tables 2.1, 3.1, 3.2, 3.3, and 3.4 can assist you in completing portions of this activity.

Table 3.1 Performance Properties of Fibers Commonly Used in Apparel

AESTHETICS AND COMFORT

FIBER	THERMAL RETENTION	ABSORPTION	RESILIENCE	DENSITY	HAND
Acetate	Poor	Fair	Poor	Medium	Excellent
Acrylic	Excellent	Poor	Good	Low	Good
Cotton	Poor	Good	Poor	High	Good
Flax	Poor	Excellent	Poor	High	Fair
Rayon	Poor	Excellent	Poor	Medium	Good
Polyester	Excellent	Poor	Good	Medium	Fair
Nylon	Good	Fair	Excellent	Low	Fair
Silk	Fair	Excellent	Fair	Low	Excellent
Spandex	Fair	Poor	Excellent	Low	Poor
Wool	Excellent	Excellent	Excellent	Medium	Good

DURABILITY AND SERVICEABILITY

FIBER	ABRASION RESISTANCE	STRENGTH	ELONGATION	RECOVERY
Acetate	Poor	Poor	Medium	Low
Acrylic	Fair	Fair	High	Medium
Cotton	Good	Good	Low	Low
Flax	Fair	Excellent	Low	Low
Rayon	Fair	Fair	Low	Medium
Polyester	Excellent	Excellent	Low	Low
Nylon	Excellent	Excellent	Low	Medium
Silk	Fair	Good	Medium	Medium
Spandex	Good	Poor	High	High
Wool	Fair	Poor	Medium	High

CONFORMANCE

FIBER	FLAMMABILITY	ODOR	TYPE OF ASH
Acetate	Burns and melts	Vinegar	Dark hard bead
Acrylic	Burns and melts	Bitter, fishy, acrid	Black hard bead
Cotton	Burns	Burning paper	Gray soft ash
Flax	Burns	Burning paper	Gray soft ash
Rayon	Burns	Burning paper	Gray soft ash
Polyester	Burns and melts	Sweet chemical	Black hard bead
Nylon	Burns and melts	Celery	Tan or dark hard bead
Silk	Burns	Burning hair	Black crisp ash
Spandex	Burns and melts	Chemical	Black soft ash
Wool	Burns	Burning hair	Black crisp ash

Sources: Allen C. Cohen and Ingrid Johnson, *J.J. Pizzuto's Fabric Science*, 10th ed. (New York: Fairchild Books, 2011), 25, 31, 36, 43, 52; Sara J. Kadolph, *Textiles*, 11th ed. (Upper Saddle River, NJ: Prentice Hall, 2010), 28, 30–31, 35.

Table 3.2 Fabric Names by Weave Structure

PLAIN WEAVE STRUCTURE FABRICS					
Batiste	Bengaline	Broadcloth	Burlap	Calico	Canvas
Challis	Chambray	Cheesecloth	Chiffon	Chintz	Crepe
Crepe de chine	Crinoline	Dotted Swiss	Duck	Faille	Flannel (can also be constructed with a twill weave)
Gauze	Georgette	Gingham	Hopsack (basket weave variation of plain weave)	Madras	Moiré
Monks cloth (basket weave variation of plain weave)	Muslin	Organdy	Organza	Ottoman	Oxford cloth (basket weave variation of plain weave)
Percale	Plissé	Pongee	Ripstop	Sailcloth	Shantung
Sheeting	Taffeta	Toile	Voile		
TWILL WEAVE STRUCTURE FABRICS					
Cavalry twill	Chino	Denim	Drill	Flannel (can also be constructed with a plain weave)	Gabardine
Herringbone	Hounds tooth	Serge	Sharkskin	Surah	Whipcord
SATIN WEAVE STRUCTURE FABRICS					
Antique satin	Brushed-back satin	Charmeuse	Crepe-back satin	Duchess satin	Peau de soie
Sateen	Slipper satin				
DOBBY WEAVE STRUCTURE FABRICS					
Birdseye	Huckaback	Pique (can also be constructed using a jacquard weave)	Shirting madras	Waffle cloth	
DOUBLE CLOTH WEAVE STRUCTURE FABRICS					
Double cloth	Kersey	Matelassé	Melton	Velvet	
JACQUARD WEAVE STRUCTURE FABRICS					
Brocade	Brocatelle	Damask	Piques (can also be constructed using a dobby weave)	Tapestry	
PILE WEAVE STRUCTURE FABRICS					
Corduroy	Terrycloth	Velveteen			

Source: Janace Bubonia, *Fashion Production Terms and Processes* (New York: Fairchild Books, 2012).

Table 3.3 Fabric Names by Knit Structure

WEFT KNIT STRUCTURE FABRICS	WARP KNIT STRUCTURE FABRICS
Balbridgan	Bobbinet
Bourrelet	Crochet
Cable knit	Filet
Double knit	Gossamer Faux fur
Fleece (weft-insertion)	Illustion
French terry (weft-insertion)	Intarsia
Interlock	Lace
Jacquard jersey	Mali (warp-insertion)
Jersey knit	Malimo (warp-insertion)
Knit terry	Mesh
Lacoste	Milanease
Lisle	Point d' esprit
Matte jersey	Pointelle
Pile jersey	Polar fleece
Piqué	Power net
Pont di Rome	Raschel knit
Purl knit	Simplex
Rib knit	Thermal knit
Velour	Tricot knit
	Tulle

Source: Janace Bubonia, *Fashion Production Terms and Processes* (New York: Fairchild Books, 2012).

Table 3.4 Yarn Number Systems

DIRECT YARN NUMBER SYSTEMS		
SYSTEM	**MASS PER UNIT LENGTH**	**FIBER TYPE**
Denier (den)	grams/9,000 meters	Filament yarns from silk or manufactured fiber
Tex (tex)	grams/1,000 meters	Spun yarns from silk or manufactured fiber
Kilotex (ktex)	kilograms/1,000 meters	Thick spun yarns from silk or manufactured fiber
Decitex (dtex)	grams/10,000 meters	Fine filament yarns from silk or manufactured fiber
INDIRECT YARN NUMBER SYSTEMS		
SYSTEM	**LINEAR DENSITY IN LENGTH PER UNIT MASS**	**FIBER TYPE**
Cotton count (cc or Ne_c)	840 yd lengths/lb	Cotton and cotton blend yarns
Linen lea (lea)	300 yd lengths/lb	Linen and linen-like yarns
Metric count (mc or Nm)	1,000 meters/gram	All spun yarns
Woolen run (wr or Nw_e)	1,600 yd lengths/lb	Coarse woolen and woolen blend yarns
Worsted count (wc or Nw_w)	560 yd lengths/lb	Fine worsted wool, worsted wool blends, and acrylic yarns

Sources: ASTM International, *Annual Book of Standards 2011: D1244-98 (2005) Standard Practice for Designation of Yarn Construction, Section Seven Textiles*, vol. 07.01 (West Conshohocken, PA: ASTM International, 2011), 262–266; ASTM International, *Annual Book of Standards 2011: D2260-03 Standard Tables of Conversion Factors and Equivalent Yarn Numbers Measured in Various Numbering Systems, Section Seven Textiles*, vol. 07.01 (West Conshohocken, PA: ASTM International, 2011), 571–583; ISO, *ISO Standard 2947 Textiles-Integrated Conversion Table for Replacing Traditional Yarn Numbers by Rounded Numbers in the Tex System* (Geneva, CH: ISO, 1973), 1–13; http://www.nexisfibers.com/spip.php?rubrique76.

OVERVIEW OF LABORATORY TESTING AND SAFETY

Lab Activities 3.2, 3.3, and 3.4 will focus on characterization testing and on the preliminary preparation for appearance and performance testing, which will require the that activities be completed over the course of several lab sessions. Lab Activities 3.2, 3.3, and 3.4 will require you to work in your Comparison Project teams of four to five students.

Lab Activity 3.2: Specimen Templates and Sampling Plan

For Lab Activity 3.2, work in your Comparison Project teams of four to five people. You will prepare specimen templates and develop a Sampling Plan for piece goods testing. For this lab, you will need 2 yards or 2 meters of woven fabric of a medium to dark color, card stock (four pages) or one package of Visi-GRID Quilter's Template Sheets (contains four plastic grid sheets, 8 1/2 inches by 11 inches) or metric equivalent, ruler, scissors, and graph paper. In the interest of time, you should divide your teams into two groups—one to work on the specimen templates and the other group to prepare the Sampling Plan. Swap your groups to check the other group's work prior to turning it in.

Preparation of Specimen Templates

Each lab group will create one set of templates for cutting specimens for testing using ASTM International (ASTM) and American Association of Textile Chemists and Colorists (AATCC) standard test methods (see specimen templates that follow). If you are using different standards from other organizations, you will need to prepare templates according to their specifications. The templates must be cut accurately and precisely from the card stock or the Visi-GRID Quilter's Template Sheets. Each lab group will prepare and turn in one set of templates. In pencil, write on the templates what test will be performed, the number of specimens that will be cut (indicate warp and filling when applicable), the grainline in the warp direction from edge to edge (on the long side of the template or on a 45-degree angle for bias), and the name of each lab group member. If templates are improperly or inaccurately cut or marked, the lab group will be expected to redo each template in question during the next lab session. It is of the utmost importance for the templates to be accurately prepared so testing results will not be compromised. The testing templates and worksheet 3.2 should be placed in an envelope when completed. Do not fold or crease the envelope or specimen templates. Some templates will be used for both piece goods testing (2-yard or 2-meter sample of fabric/lab sample) labs and Comparison Project lab activities, whereas others will be used for one or the other.

Your instructor may require preparation of templates that are slightly different from the ones that follow, depending upon the testing equipment and apparatuses in your lab and the available test standards. Only one template should be prepared for each of the following tests. Templates will be used to mark and cut specimens on the piece goods lab sample or the Comparison Project garment designated for physical testing.

Piece Goods Lab Specimen Templates

- Fabric Weight—6-inch square or 15-centimeter square (used for piece goods lab only)—cut three round specimens (4 7/16-inch diameter or 11.3-centimeter diameter) from these squares with a special fabric sample cutter for testing fabric weight.
- Crocking—2 inches wide by 5.1 inches long or 50 millimeters wide by 130 millimeters long—*mark grain on template in bias direction (used for both labs)—cut two.
- Tensile Strength [Grab Test—9.2 Grab Test G]—4 inches wide by 6 inches long or 100 millimeters wide by 150 millimeters long (piece goods lab only)—cut five warp, eight filling.
- Tearing Strength by the Tongue Procedure (Single Rip) —3 inches wide by 8 inches long or 75 millimeters wide by 200 millimeters long (piece goods lab)—cut five warp, five filling.
- Abrasion Resistance (Rotary Platform, Double Head Method)—6-inch square or 15-centimeter square (piece goods lab)—cut five (three to test and two for control specimens).

Comparison Project Lab Specimen Templates

- Fabric Weight—Cut three 4 7/16-inch diameter or 11.3-centimeter diameter round specimens directly from garment using the special fabric cutter for testing fabric weight—do not cut 6-inch square or 15-centimeter square first.
- Crocking—2 inches wide by 5.1 inches long or 50 millimeters wide by 130 millimeters long—*mark grain on template in bias direction (used for both labs).
- Pilling Resistance (Random Tumble Pilling)—4.13-inch square or 105-millimeter square—*mark grain on template in bias direction (Comparison Project lab only)—cut three.
- Bursting Strength (Ball Burst)—5-inch square or 125-millimeer square (Comparison Project lab only)—cut five if possible or a minimum of three if there is not enough available garment material.

Specimen Templates

Use the following to assist with preparing the specimen templates on the Visi-GRID Quilter's Template sheets or card stock.

Fabric Weight

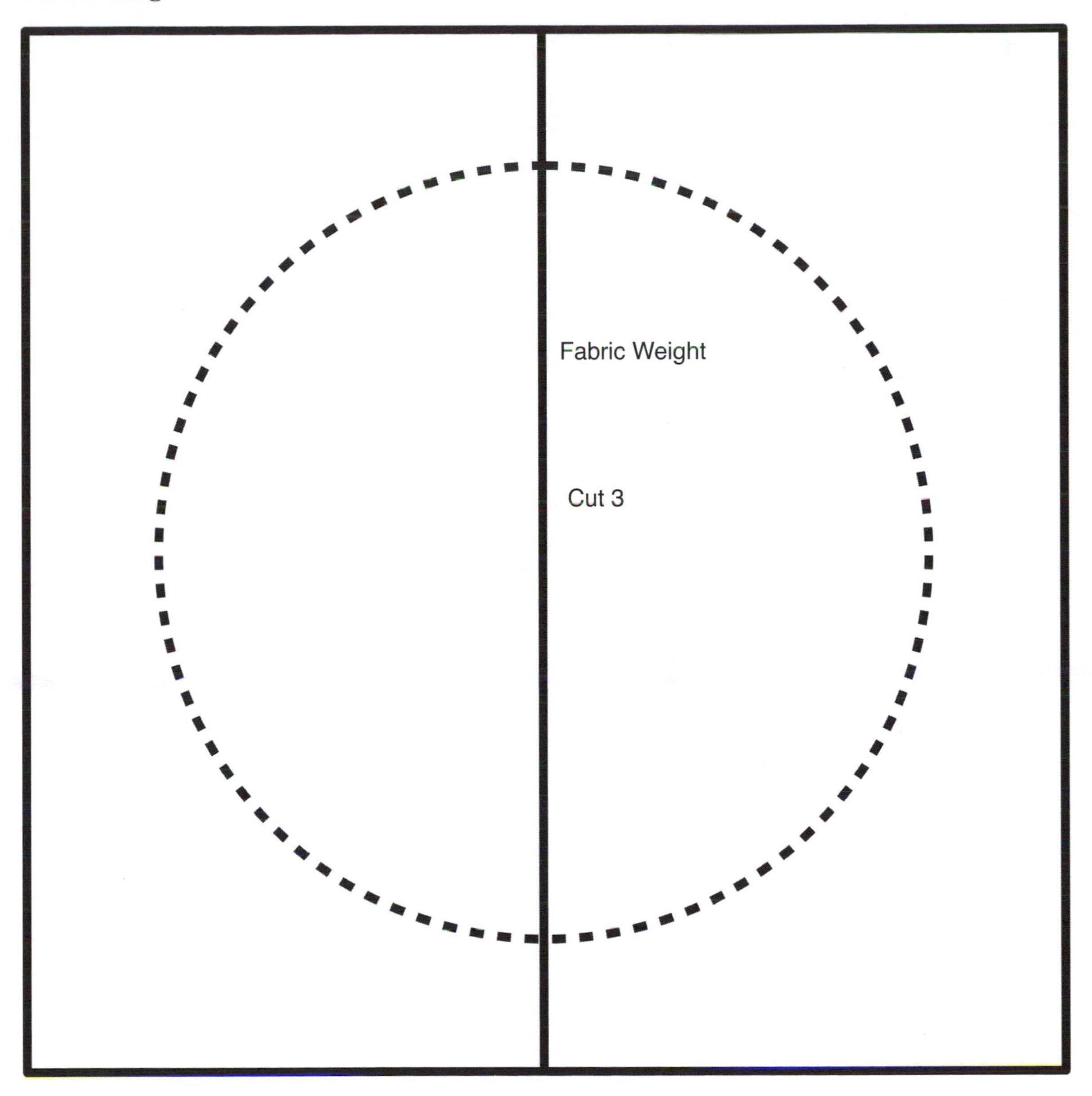

Crocking Template

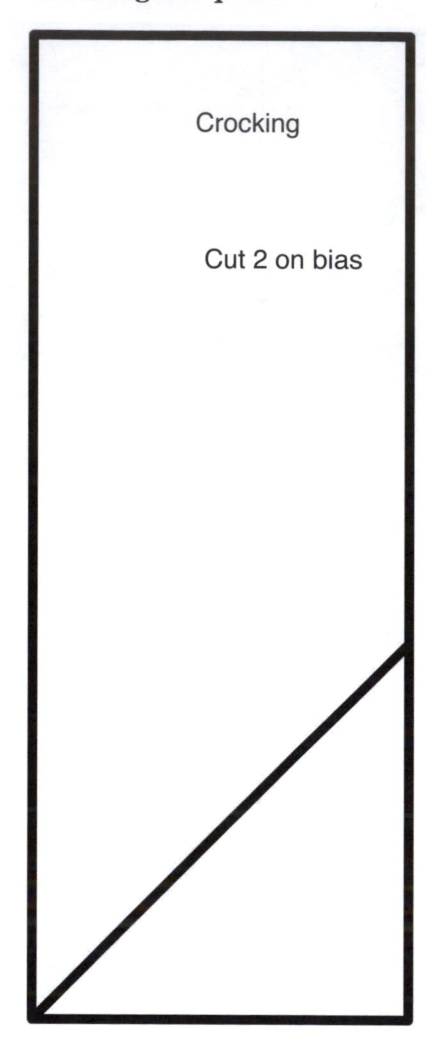

Tensile Strength Template—Grab Test

	Tensile Strength - Grab Test Cut 5 warp and 8 filling

Tearing Strength Template—Tongue Procedure

Abrasion Resistance Template

	Abrasion Resistance Cut 5

Pilling Resistance Template—Random Tumble

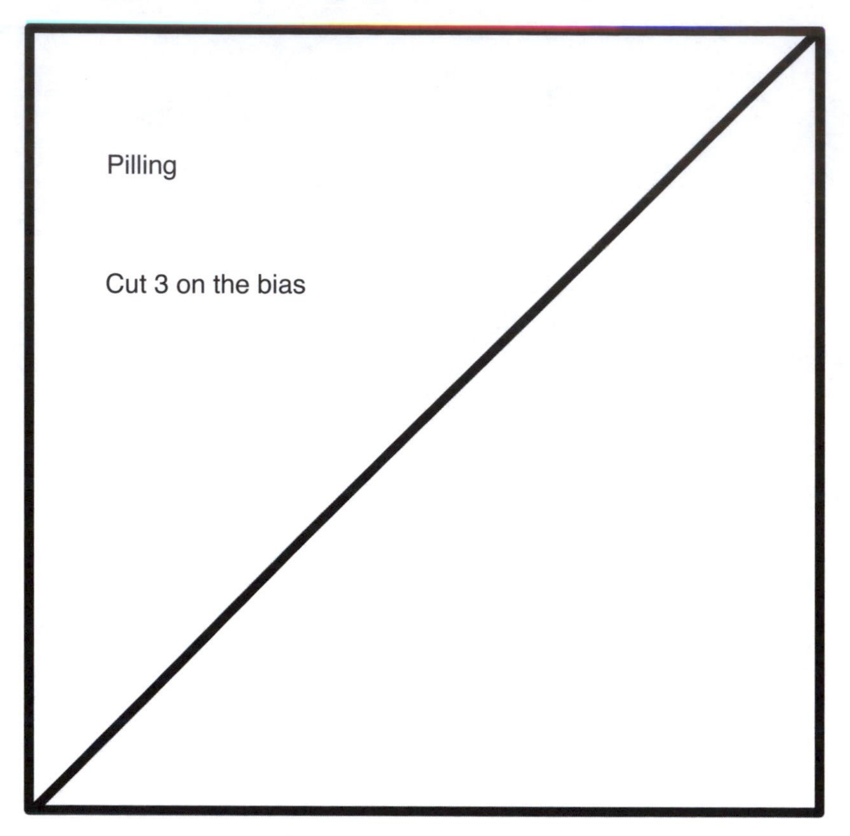

Bursting Strength Template—Ball Burst

	Bursting Strength Cut 5

Preparation of Sample Plan

A sample plan is a map of where specimens will be cut on the 2-yard/2-meter lab sample. You will create a sample plan based on the width of the fabric purchased. Read the directions completely before beginning. Each lab group will create one Sampling Plan for taking specimens from their piece goods lab sample. The Sampling Plan will be mapped out on the graph paper. When mapping specimens, it is important to place the templates close together to utilize the sample wisely. Template edges can touch (see Figure 3.1).

Figure 3.1

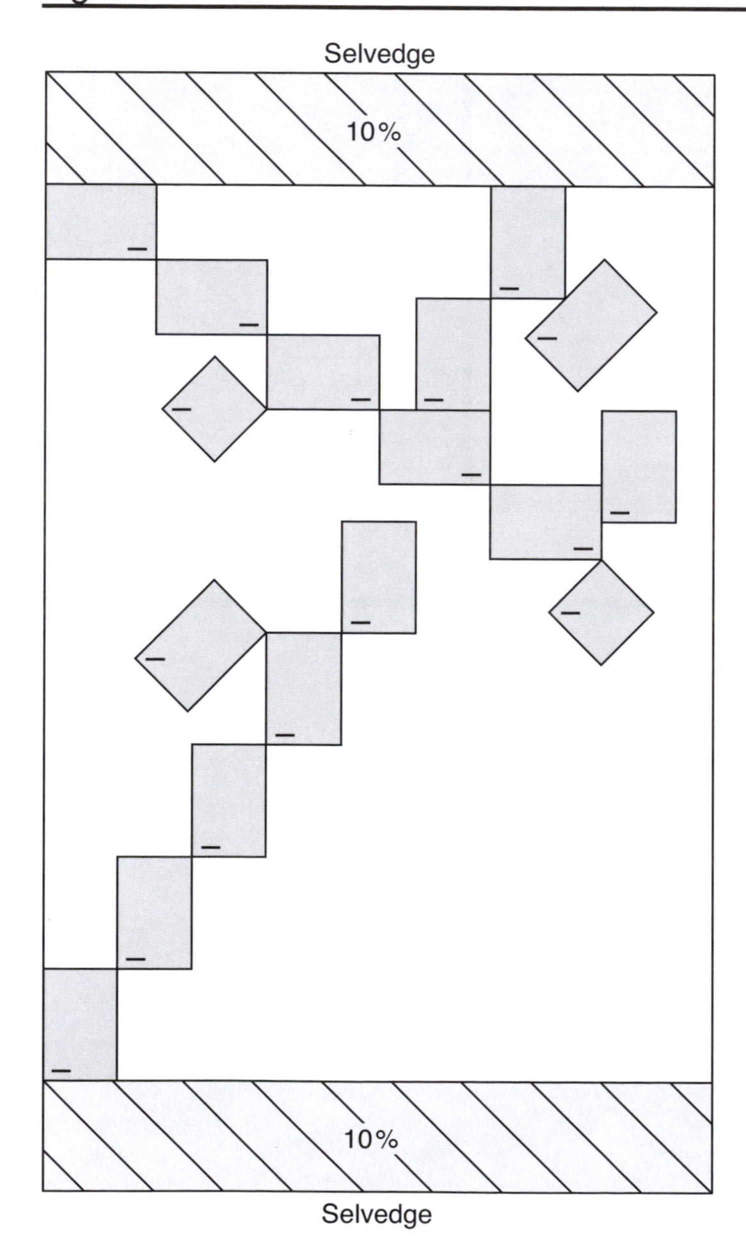

Rules for Taking Specimens

1. Follow the test method directions for the designating number of specimens to cut and how to prepare them.
2. Do not lay out or cut any specimen closer to the fabric selvage than 10 percent of the fabric width.
3. Lay out and cut specimens for the same test by staggering them in different areas of the lab sample to capture different warp and filling yarns within each specimen.
4. Mark and cut specimens accurately, exactly, and on-grain or on-bias as specified by the test method.
5. Identify warp direction of each specimen by marking a line | or ↑ arrow.
6. It is important to keep warp marks on each specimen small (1/2 inch is adequate) and away from critical test areas on each specimen. Mark the warp in the lower right or left corner of the specimen close to the cut edge. Do not mark the warp grainline edge to edge on fabric specimens.
7. Do not lay out or cut specimens containing seams, trim, buttons, buttonholes, or attached decorative elements unless the specified element is to be tested (e.g., if you are testing for seam strength, seam slippage, button impact).

The group will begin by measuring the lab sample and drawing it to scale on the graph paper, 1 inch or 1 centimeter being equal to one square on the graph paper. Mark the cut edges as "raw edge" and the selvedges as "selvedge." Measure in 10 percent of the fabric width from both selvages, and draw lines that extend the length of the fabric. No specimens can be placed within these areas. For example, if your fabric is 60 inches wide or 152 centimeters, you will measure in 6 inches or 15.2 centimeters from each selvage. Then mark the warp direction with a small | or ↑.

Once the outline of the lab sample is drawn to scale on the graph paper, the group members will determine how each test specimen will be placed on the sample. Begin with the greatest amount of specimens that will need to be cut and proceed to the least amount needed. After you have determined the placement of the test specimens, draw them to scale on the Sampling Plan. Create a key (i.e., FW = fabric weight, TSG = tensile strength grab method, TST = tensile strength tongue method, AR = abrasion resistance, and CR = crocking), and mark the specimens with the abbreviation that corresponds with the key. Number each of the specimens to be cut (i.e., TSG 1W; TSG 2W—meaning, respectively, tensile strength grab method, specimen 1 warp direction; tensile strength grab method, specimen 2, warp direction). Make sure you have marked the warp direction on all specimens with a small | or ↑ in the lower left or right corner.

When the Sampling Plan is complete, each group should have a sheet of graph paper with a drawing of the piece goods lab sample indicating where all of the test specimens will be cut. Check the plan to be sure that the specimen templates are placed on the correct grain (on-grain or on-bias) depending upon the tests to be conducted, and that the warp direction is marked on all specimens. Make sure the specimens are staggered for each test so as many shuttle changes can be utilized as possible. Try to utilize the sample area so wasted fabric is at a minimum. All of your specimens must be cut from one lab sample.

Each group will turn in one Sampling Plan. All lab group members' names must appear on the Sampling Plan. To ensure you have completed the specimen templates and Sampling Plan activity accurately and completely, use Lab Activity 3.2: Specimen Templates and Sampling Plan to verify this. After you have completed the templates and Sampling Plan and your instructor has approved them, begin marking your piece goods lab sample (2-yard/2-meter sample) according to the Sampling Plan. Begin cutting and preparing the specimens for testing in a future lab.

Lab Activities 3.3 and 3.4: Characterization Testing

Characterization testing on raw materials is discussed in Chapter 11 of *Apparel Quality: A Guide to Evaluating Sewn Products*. It is important to begin some of these tests early in the term to allow enough time for testing and evaluation to occur. The recommended characterization tests for this lab include fiber identification, yarn construction, fabric count, and weight. Characterization testing will be performed on the piece goods lab sample and recorded on worksheet 3.3. Follow the Sampling Plan when cutting weight specimens. Mount the yarn specimens and swatch of fabric in the designated areas of worksheet 3.3. Specimens for fabric weight should be mounted on cardstock and attached to worksheet 3.3.

Worksheet 3.4 will be used to record characterization test result for your Comparison Project garment. For the Comparison Project you have three garments, one for wear-testing, one for physical testing, and one for a control. Use the garment designated for physical testing when marking and cutting specimens for characterization testing. When cutting weight specimens from the Comparison Project garment make sure you have planned where the other test specimens will be taken from. Refer to Table 3.5 for Common Standard International Test Methods for Fabric Characterization and Table 3.6 for Standard Conditioning Times for Testing Textiles for Apparel. Mount the garment labels, yarn specimen, and swatch of fabric in the designated areas of worksheet 3.4. Specimens for fabric weight should be mounted on cardstock and attached to worksheet 3.4.

Table 3.5 Common Standard International Test Methods for Fabric Characterization

FIBER IDENTIFICATION	
AATCC 20	Fiber Analysis: Qualitative
AATCC 20A	Fiber Analysis: Quantitative
ASTM D276	Standard Test Method for Identification of Fibers in Textiles
BS EN ISO 1833-10	Textiles Quantitative Chemical Analysis. Mixtures of Triacetate or Polylactide and Certain Other Fibres (Method Using Dichloromethane)
BS EN ISO 1833-11	Textiles Quantitative Chemical Analysis. Mixtures of Cellulose and Polyester Fibres (Method Using Sulfuric Acid)
BS EN ISO 1833-12	Textiles Quantitative Chemical Analysis. Mixtures of Acrylic, Certain Modacrylics, Certain Chlorofibres, Certain Elastanes and Certain Other Fibres (Method Using Dimmethylformamide)
BS EN ISO 1833-13	Textiles Quantitative Chemical Analysis. Mixtures of Chlorofibres and Certain Other Fibres (Method Using Carbon Disulfide/Acetone)
BS EN ISO 1833-14	Textiles Quantitative Chemical Analysis. Mixtures of Acetate and Certain Chlorofibres (Method Using Acetic Acid)
BS EN ISO 1833-15	Textiles Quantitative Chemical Analysis. Mixtures of Jute and Certain Animal Fibres (Method by Determining Nirtogen Content)
BS EN ISO 1833-16	Textiles Quantitative Chemical Analysis. Mixtures of Polypropylene Fibres and Certain Other Fibres (Method Using Xylene)
BS EN ISO 1833-17	Textiles Quantitative Chemical Analysis. Mixtures of Chlorofibres (Homopolymers of Vinyl Chloride) and Certain Other Fibres (Method Using Sulfuric Acid)
BS EN ISO 1833-18	Textiles Quantitative Chemical Analysis. Mixtures of Silk and Wool or Hair (Method Using Sulfuric Acid)
BS EN ISO 1833-20	Textiles Quantitative Chemical Analysis. Mixtures of Elastane and Certain Other Fibers (Method Using Dimethylacetamide)
BS EN ISO 1833-21	Textiles Quantitative Chemical Analysis. Mixtures of Chlorofibres, Certain Modacrylics, Certain Elastanes, Acetates, Triacetates and Certain Other Fibres (Method Using Cycloheanone)
BS EN ISO 1833-22	Textiles Quantitative Chemical Analysis. Mixtures of Visscose or Certain Types of Cupro or Modal or Lyocell and Flax Fibres (Method Using Formic Acid and Zinc Chloride)
BS EN ISO 1833-24	Textiles Quantitative Chemical Analysis. Mixtures of Polyester and Certain Fibres (Method Using Phenol and Tetraachloroethane)
BS EN ISO 1833-26	Textiles Quantitative Chemical Analysis. Mixtures of Melamine and Cotton or Aramide Fibres (Method Using Hot Formic Acid)
BS ISO 2076	Textiles—Man-made Fibres. Generic Names
BS ISO17751	Textiles—Quantitative Analysis of Animal Fibers by Microscopy—Cashmere, Wool, Specialty Fibres and Their Blends
JIS L 1030-1	Test Methods for Quantitative Analysis of Fiber Mixtures of Textiles—Part 1: Testing Methods for Fiber Identification
JIS L 1030-2	Test Methods for Quantitative Analysis of Fibre Mixtures of Textiles—Part 2: Testing Methods for Quantitative Analysis of Fibre /Mixtures

continued

Table 3.5 *(cont.)*

YARN CONSTRUCTION	
ASTM D861	Practice for Use of the Tex System to Designate Linear Density of Fibers, Yarn Intermediates, and Yarns
ASTM D1244	Standard Practice for Designation of Yarn Construction
BS EN ISO 2061	Textiles—Determination of Twist in Yarns—Direct Counting Method
ISO 2	Textiles—Designation of the Direction of Twist in Yarns and Related Products
ISO 1139	Designation of Yarns
ISO 1144	Textiles—Universal System for Designated Linear Density (Tex System)
JIS L 0101	Tex System to Designate Linear Density of Fibres, Yarn Intermediate Yarns and Other Textile Materials
FABRIC THICKNESS	
ASTM 1777	Standard Test Method for Thickness of Textile Materials
BS EN ISO 5084	Textiles—Determination of Thickness of Textiles and Textile Products
FABRIC WEIGHT	
ASTM D3776/3776M	Standard Test Methods for Mass Per Unit Area (Weight) of Fabric
ASTM D3887	Standard Specification for Tolerances for Knitted Fabrics: Section 9 Test Method—Weight (Mass)
ISO 3801	Textiles—Woven Fabrics—Determination of Mass Per Unit Length and Mass Per Unit
FABRIC COUNT	
ASTM D3775	Standard Test Method for Warp (End) and Filling (Pick) Count of Woven Fabrics
ASTM D3887	Standard Specification for Tolerances for Knitted Fabrics: Section 12 Test Method—Fabric Count
ISO 7211-1	Textiles—Woven Fabrics—Construction—Methods of Analysis Part 1: Methods for the Presentation of a Weave Diagram and Plans for Drafting, Denting, and Lifting
ISO 7211-2	Textiles—Woven Fabrics—Construction—Methods of Analysis Part 2: Determination of Number of Threads per Unit Length
ISO 7211-3	Textiles—Woven Fabrics—Construction—Methods of Analysis Part 3: Determination of Crimp of Yarn in Fabric
ISO 7211-4	Textiles—Woven Fabrics—Construction—Methods of Analysis Part 4: Determination of Twist in Yarn Removed from Fabric
ISO 7211-5	Textiles—Woven Fabrics—Construction—Methods of Analysis Part 5: Determination of the Linear Density of Yarn Removed from Fabric
ISO 7211-6	Textiles—Woven Fabrics—Construction—Methods of Analysis Part 6: Determination of the Mass of Warp and Weft per Unit Area of Fabric

Table 3.6 Standard Conditioning Times for Testing Textiles for Apparel

FIBER CONTENT	MINIMUM CONDITIONING TIME REQUIRED (IN HOURS)
Animal and protein fibers	8
Cellulose fibers	6
Viscose	8
Acetate	4
Manufactured fibers	2
Blended fibers	The fiber component within the blend that requires the longest conditioning period shall be used when determining the minimum conditioning time required.

Sources: ASTM International, "ASTM D1776-08e1 Standard Practice for Conditioning and Testing Textiles," West Conshohocken, PA: ASTM International, 2014; British Standards Institute, BS EN ISO 139: 2005 + A1: 2011, *Textiles Standard Atmosphere for Conditioning and Testing*, Brussels, BE: CEN European Committee for Standardization.

Lab Activity 3.5: Wear-Testing Comparison Project Garment

Measure the Comparison Project garment that you will wear, following test method AATCC 150 Dimensional Changes of Garments after Home Laundering, ASTM D6321 Standard Practice for the Evaluation of Machine Washable T-Shirts, or an equivalent standard, in preparation for the wear test that will begin this week. ASTM D3181 Standard Guide for Conducting Wear Tests on Textiles or an equivalent standard will be used. Mark the comparison garment that you will wear with benchmarks for testing garment twist following test method AATCC 179 Skewness Change in Fabric and Garment Twist Resulting from Automatic Home Laundering, using Method 2: inverted T marking.

After measuring the comparison garment and marking benchmarks for determining skewness, put on your shirt and take a "before" photo. This must be done before any testing begins. Make sure to note where the photo was taken, the lighting condition, and the distance from the camera because you will need this information when taking your "after" photo. This is critical because the before and after photos will be used in your comparison and performance evaluation.

Once the original measurements are taken according to the standard you are using, and the "before" photo is taken, wear the garment one time for 4 hours minimum and 6 hours maximum and refurbish it before the next lab session. Every lab session for the next 5 weeks will require you to wear the garment, refurbish it, and bring it to lab to complete dimensional change and further testing.

Lab Activity 3.1: RAW MATERIALS CLASSIFICATION

Identify the illustrations of fibers, yarns, and fabric structure classifications. List the basic knit and weave classifications, and provide examples of fabric names for each. Answer the questions regarding raw materials.

If your lab meets OSHA regulations for conducting flammability testing, you will burn yarns to identify fiber type, flammability, odor, and type of ash. Do not burn yarns outside of lab or in a lab that does not meet OSHA regulations for this type of activity.

Your Name

Name(s) of Teammate(s)

Fiber Identification	
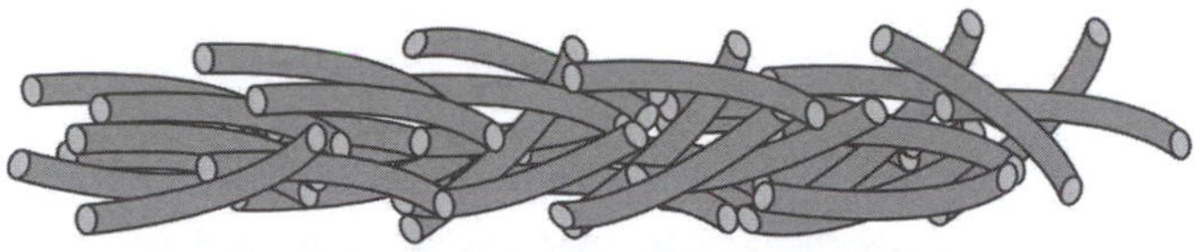	
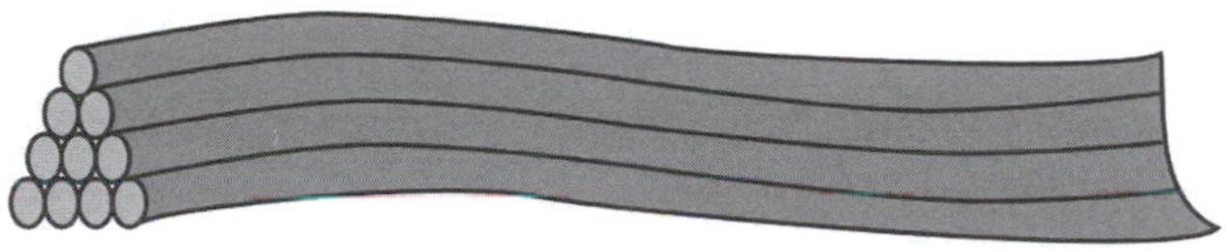	

Yarn Construction	

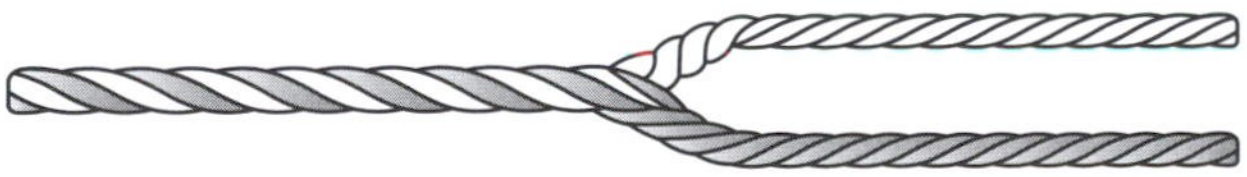	

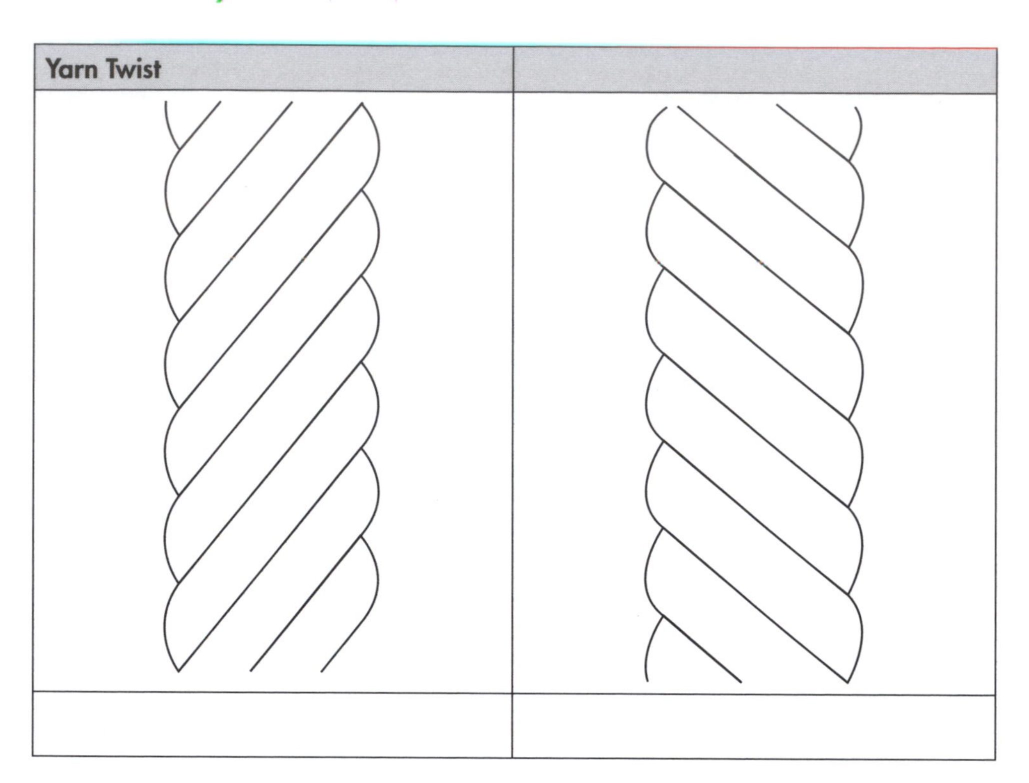

BURN TESTING TO DETERMINE FIBER TYPE

Your instructor will provide five yarns for you to burn to determine fiber type, flammability, odor, and type of ash. Mount remaining portions of yarns in the Fiber Type/Name boxes under the name of the fiber. See Table 3.1. *Do not burn yarns outside of lab or in a lab that does not meet OSHA regulations for this type of activity.*

Fiber Type/Name	Flammability	Odor	Type of Ash

Lab Activity 3.1 *(cont.)*

For each basic weave type, list three examples of fabrics (use appropriate names) that utilize the specified weave structure. See Table 3.2.

Basic Weave Structures

Plain Weave	Twill Weave	Satin Weave

Select three fabrics from the preceding list, and indicate the type of garment for which the material could be used. Consider the characteristics of the material and why it would be appropriate for use in the designated garment.

Fabric Name	Garment Type

For each basic knit type, list three examples of fabrics (use appropriate names) that utilize the specified knit structure. See Table 3.3.

Basic Knit Structures

Weft Knits	Warp Knits

Select two fabrics from the preceding list, and indicate the type of garment for which the material could be used. Consider the characteristics of the material and why it would be appropriate for use in the designated garment.

Fabric Name	Garment Type

Lab Activity 3.1 *(cont.)*

List three ways of determining whether a fabric is woven or knitted, by using visual examination and touch only.

1.
2.
3.

Match the Yarn Number System with the yarn number designations. See Table 3.4.
A. Direct Yarn Number System *B. Indirect Yarn Number System*

________ Cotton count (cc or Ne_c)
________ Decitex (dtex)
________ Denier (den)
________ Kilotex (ktex)
________ Linen Lea (lea)
________ Metric count (mc or Mn)
________ Tex (tex)
________ Woolen run (wr or Nw_e)
________ Worsted count (wc or NW_w)
________ Finer the yarn the lower the yarn number
________ Finer the yarn the higher the yarn number

List three aesthetic finishes and three functional finishes and the desired effect from application. Provide one example of a garment type/style where each aesthetic or functional finish would be beneficial. See Table 2.1.

Aesthetic Finishes	Appropriate for This Type/Style of Garment
Functional Finishe	**Appropriate for this Type/Style of Garment**

Lab Activity 3.2: SPECIMEN TEMPLATES AND SAMPLING PLAN

Use the following checklists as a means to ensure you have prepared the specimen templates and Sampling Plan correctly and completely.

Your Name

Name(s) of Teammate(s)

Specimen Templates Prepared

Template Name	Template Dimensions	Grain Direction	Contains Required Labeling

Lab Activity 3.2 *(cont.)*

SAMPLING PLAN PREPARED

Carefully review the Sampling Plan and instructions to make sure all of the necessary specimens and information are documented correctly. Answer the following questions. If you indicate No for any, make the necessary corrections before turning in the Sampling Plan.

Sampling Plan Questions	Answer Yes or No
Did you mark the raw edges and selvedges on the Sampling Plan?	
Did you measure in 10 percent of the fabric width from both selvages and avoid placing any specimens in these areas?	
Does the Sampling Plan contain all of the needed specimens for all of the tests to be conducted?	
Did you mark the warp direction on all specimens?	
Did you label each specimen with the correct test code and did you include a key?	

Test Name	Number of Warp Specimens	Number of Filling Specimens	Grain Direction	Contains Required Labeling

Lab Activity 3.3: CHARACTERIZATION TESTING ON PIECE GOODS

Your Name

Name(s) of Teammate(s)

Before beginning any tests, record the temperature and relative humidity in your lab (if available).

Fabric Name	Basic Weave Structure	Fiber Content
Color(s)	**Fabric Width**	**Purchase Price per Yard/Meter**

If the test methods you are using are different from the following methods or if additional tests are conducted, use the empty grid to write in the standards you are using and the data collected.

Test Method	Name of Test	Report	Results	Results	°F °C %RH
ASTM D1244	Standard Practice for Designation of Yarn Construction	Report yarn #, system, twist, tpi Ex. 24 Ne_c S 16 tpi or report the information you can obtain	______ Mount warp yarn here	______ Mount filling yarn here	
ASTM D3775	Standard Test Method for Warp (End) and Filling (Pick) Count of Woven Fabrics	Count = warp by filling (Report count and mean to the nearest whole number)	Warp 1. 2. 3. 4. 5. Warp Mean ______	Filling 1. 2. 3. 4. 5. Filling Mean ______	

Lab Activity 3.3 *(cont.)*

Test Method	Name of Test	Report	Results	Results	°F °C %RH
ASTM 3776/3776M	Standard Test Methods for Mass Per Unit Area (Weight) of Fabric	Mass in ounces per yard^2 and grams per meter^2	Weight in grams per meter^2 1. 2. 3. _______Mean	Weight in ounces per yard^2 1. 2. 3. _______Mean	
Test Method	**Name of Test**	**Report**	**Results**	**Results**	**°F °C %RH**
Test Method	**Name of Test**	**Report**	**Results**	**Results**	**°F °C %RH**
Test Method	**Name of Test**	**Report**	**Results**	**Results**	**°F °C %RH**

Lab Activity 3.3 *(cont.)*

Test Method	Name of Test	Report	Results	Results	°F °C %RH

Test Method	Name of Test	Report	Results	Results	°F °C %RH

Mount fabric swatch here

Lab Activity 3.4: CHARACTERIZATION TESTING ON COMPARISON PROJECT GARMENT

Your Name

Name(s) of Teammate(s)

Before beginning any tests, record the temperature and relative humidity in your lab (if available).

Brand	**Retailer**	**Original Price**
Fabric Name	**Basic Knit Structure**	**Purchase Price**
Color(s)	**Country of Origin**	**Size**
Garment Description	**Care Label**	**Fiber Content**
	Mount label here	Mount label here

If the test methods you are using are different from the following methods, use the empty grid to write in the standards you are using and the data collected.

Test Method	**Name of Test**	**Report**	**Results**		**°F °C %RH**
ASTM D1244	Standard Practice for Designation of Yarn Construction	Report yarn #, system, twist, tpi Ex. 24 Ne_c S 16 tpi or report the information you can obtain	Mount yarn here		

Lab Activity 3.4 *(cont.)*

Test Method	Name of Test	Report	Results	Results	°F °C %RH
ASTM D3887	Standard Specification for Tolerances of Knitted Fabrics See section 12. Test Method—Fabric Count	Count = wales by courses (Report count and mean to the nearest whole number)	Wales 1. 2. 3. 4. 5. Wales mean __________	Courses 1. 2. 3. 4. 5. Courses mean __________	
ASTM 3776/3776M	Standard Test Methods for Mass Per Unit Area (Weight) of Fabric	Mass in ounces per yard^2 and grams per meter^2	Weight in grams per meter^2 1. 2. 3. ______Mean	Weight in ounces per yard^2 1. 2. 3. ______Mean	
Test Method	**Name of Test**	**Report**	**Results**	**Results**	**°F °C %RH**
Test Method	**Name of Test**	**Report**	**Results**	**Results**	**°F °C %RH**

Lab Activity 3.4 *(cont.)*

Test Method	Name of Test	Report	Results	Results	°F °C %RH
Test Method	**Name of Test**	**Report**	**Results**	**Results**	**°F °C %RH**
Test Method	**Name of Test**	**Report**	**Results**	**Results**	**°F °C %RH**
Test Method	**Name of Test**	**Report**	**Results**	**Results**	**°F °C %RH**

Lab Activity 3.5: WEAR-TESTING COMPARISON PROJECT GARMENT

Your Name

Name(s) of Teammate(s)

Record information pertaining to garment fit and appearance.

Garment Comfort and Fit of Garment by Location (i.e., arm hole, neckline, collar, waistband, length, width)	**Original Condition Areas That Do Not Lay Smoothly** (i.e., gap, bulge, wrinkle, unusual folds, tightness, looseness)	**Condition After Fifth Wear and Refurbishment Areas That Do Not Lay Smoothly** (i.e., gap, bulge, wrinkle, unusual folds)
Garment Location	**Problem Areas**	**Problem Areas**

Lab Activity 3.5 *(cont.)*

Calculate the cost per wear for your garment based on wearing it five times.

Purchase Price	Number of Times Worn	Cost Per Wear (Purchase price ÷ Number of times garment was worn = Cost per wear)

Document when you wear the garment and when it is cleaned.

Wear Log Date Worn	Start Time (hours and minutes)	End Time (hours and minutes)	Date Refurbished
Week 1			
Week 2			
Week 3			
Week 4			
Week 5			

Document how you refurbished the garment for each of the 5 weeks of wear-testing.

Care Method Followed (as stated on garment care label)				
Home Laundering	**Detergent and Amount Was fabric softener, dryer sheet, or bleach used?**	**Drying Time (note if not dry within 30 minutes)**	**Additional Drying Time (if applicable – monitor every 15 minutes)**	**Iron Setting (if applicable)**
Week 1	Detergent: ☐1/4 cup ☐1/2 cup ☐3/4 cup ☐1 cup ☐Fabric softener ☐Dryer sheet ☐Nonchlorine bleach	☐Dry ☐Not dry	☐15 minutes more ☐30 minutes more	☐Low ☐Medium ☐High ☐No steam

Lab Activity 3.5 *(cont.)*

Week 2	Detergent: ☐1/4 cup ☐1/2 cup ☐3/4 cup ☐1 cup ☐Fabric softener ☐Dryer sheet ☐Nonchlorine bleach	☐Dry ☐Not dry	☐15 minutes more ☐30 minutes more	☐Low ☐Medium ☐High ☐No steam
Week 3	Detergent: ☐1/4 cup ☐1/2 cup ☐3/4 cup ☐1 cup ☐Fabric softener ☐Dryer sheet ☐Nonchlorine bleach	☐Dry ☐Not dry	☐15 minutes more ☐30 minutes more	☐Low ☐Medium ☐High ☐No steam
Week 4	Detergent: ☐1/4 cup ☐1/2 cup ☐3/4 cup ☐1 cup ☐Fabric softener ☐Dryer sheet ☐Nonchlorine bleach	☐Dry ☐Not dry	☐15 minutes more ☐30 minutes more	☐Low ☐Medium ☐High ☐No steam
Week 5	Detergent: ☐1/4 cup ☐1/2 cup ☐3/4 cup ☐1 cup ☐Fabric softener ☐Dryer sheet ☐Nonchlorine bleach	☐Dry ☐Not dry	☐15 minutes more ☐30 minutes more	☐Low ☐Medium ☐High ☐No steam

Test Method	**Name of Test**				**Report**			
AATCC 150	Dimensional Changes of Garments after Home Laundering				Average % DC in both length and width directions to the nearest 0.01%			
ASTM 6321	Standard Practice for the Evaluation of Machine Washable T-Shirts				Average % DC in both length and width directions to the nearest 0.01%			
Dimensional Stability	**Length**				**Width**			
	Front	Back	Sleeve	% DC Length Mean	Chest	Neck Opening	Armscye	% DC Width Mean
Original measurements								
Week 1 measurements								
Week 1 % DC								
Week 2 measurements								

Lab Activity 3.5 *(cont.)*

Week 2 % DC								
Week 3 measurements								
Week 3 % DC								
Week 4 measurements								
Week 4 % DC								
Week 5 measurements								
Week 5 % DC								

Calculation:
Measurement after wash – Original measurement ÷ Original measurement × 100 =
Dimensional change % *Calculate to nearest 0.1%*

Test Method	Name of Test	Report
AATCC 179	Skewness Change in Fabric and Garment twist Resulting from Automatic Home Laundering—Garment marking method 2, Calculation option 3	Average % change in skewness/Garment twist to the nearest 0.01% Calculation: % Change in skewness = 100 × (A A′/AB)
ASTM 6321	Standard Practice for the Evaluation of Machine Washable T-Shirts	

Lab Activity 3.5 *(cont.)*

Skewness	Measurement between A and A' A = original mark A' = offset mark	% change in skewness/ Garment twist	Indicate if skewness is to the left or to the right. Positive % change = Skewness to the left Negative % change = Skewness to the right
Week 1			□No skewness □Skewness to the left □Skewness to the right
Week 2			□No skewness □Skewness to the left □Skewness to the right
Week 3			□No skewness □Skewness to the left □Skewness to the right
Week 4			□No skewness □Skewness to the left □Skewness to the right
Week 5			□No skewness □Skewness to the left □Skewness to the right

Mount before and after photos of comparison garment here.

Original Condition

Insert Photo "Before" Here

After Fifth Wear and Refurbishment

Insert Photo "After" Here

NOTES

CHAPTER 4

Garment Construction Details Lab

LAB OBJECTIVES:

- To analyze construction details and identify shaping methods used to achieve fit in garments.
- To identify the openings, facings, closures, and hem finishes used for mass-produced apparel products.
- To understand the importance of closure selection for apparel items and ease of donning and doffing.

In Chapter 4 of *Apparel Quality: A Guide to Evaluating Sewn Products*, garment construction details and methods for achieving fit of apparel products are discussed. There are many options for shaping garments to contour the body while maintaining the appropriate amount of functional ease and aesthetic appeal. When designers make selection decisions for the style of openings, closures, and hem finishes for production, they must carefully consider the garment style, materials, functionality for intended use, care, and compatibility of all component materials. They must also consider their target market's expectations when purchasing an apparel item regarding price, quality, and value.

There are three lab activities (4.1, 4.2, 4.3) for Chapter 4 and two additional activities carried over from Chapter 3, (the piece goods preparation for Lab Activity 3.2 and the Comparison Project garment analysis after week 1 of the wear test for Lab Activity 3.5). The Chapter 4 lab activities will focus on analyzing garment construction details and shaping methods. Any lab work that has not been completed from Chapter 3 such as the preparation of templates for worksheet 3.2, the Sampling Plan, or characterization testing for worksheets 3.3 and 3.4 should be completed during this lab session. You will also mark the piece goods lab sample according to the Sampling Plan and begin cutting and preparing the specimens for testing in a future lab.

PREPARATION AND SUPPLIES FOR CHAPTER 4 LAB ACTIVITIES

In preparation for Chapter 4 lab activities each group member (small group of 2 to 3) will bring their one woven shirt from lab activity 2.1 or meet with your group and select another garment. Each garment within the group should have similar styling and be the same type of garment, i.e. dress, pants/trousers, jacket, etc. A minimum of two price categories should be represented within each two person lab group or a minimum of three price classifications for each three person lab group (budget, moderate, better, contemporary, bridge, designer). Prior to Chapter 4 lab activities, you will also need to wear the comparison project garment for the designated number of hours, clean it according to the garment label care instructions, and bring it along with the control garment to lab for evaluation. If your team has not completed preparation of templates, the sampling plan and marking and cutting of piece goods specimens the following supplies from Chapter 3 lab activities will be needed: 4 pages of card stock or 1 package of Visi-GRID Quilter's Template Sheets (contains 4 plastic grid sheets, 8 ½ inches by11inches or metric equivalent), graph paper, a permanent marking pen (such as Sharpie rub-a-dub), ruler, scissors, 2 additional pieces of card stock for mounting specimens, and 2 yard/2 meter lab sample of piece goods.

Lab Activity 4.1: Garment Analysis of Construction Details across Price Classifications

For Lab Activity 4.1: Garment Analysis of Construction Details across Price Classifications, work in teams of two to three people, Each group member will bring the woven shirt used for Lab Activity 2.1 to lab for further evaluation or select another woven garment. Each garment within the group should have similar styling and be the same type of garment (i.e., dress, pants/trousers, jacket). A minimum of two price categories (budget, moderate, better, contemporary, bridge, designer) should be represented within each two-person lab group, or a minimum of three price classifications should be available for each three-person lab group. As a team, begin completing worksheet 4.1. Discuss and record the fit and construction details, shaping methods, openings, closures, and hem finishes used. Consult Table 4.1 and figures from Chapter 4 of the text for assistance with identification and use of proper terminology. After you have completed worksheet 4.1, move on to Lab Activity 4.2: Comparison Project Garment Analysis of Construction Details across Competing Brands.

Lab Activity 4.2: Comparison Project Garment Analysis of Construction Details across Competing Brands

For Lab Activity 4.2: Comparison Project Garment Analysis of Construction Details across Competing Brands, work in your Comparison Project teams of four to five people. Each group member must bring his or her Comparison Project control garment. As a team, begin completing activity worksheet 4.2. Discuss and record the fit and construction details, shaping methods, openings, closures, and hem finishes used. Consult Table 4.1 and the figures from Chapter 4 of the text for assistance with identification and use of proper terminology. After you have completed worksheet 4.2, move on to the wear-testing Comparison Project garment.

Table 4.1 Button Sizes

LIGNE NUMBER	BUTTON DIAMETER
14L	5/16 inch (8 mm)
16L	3/8 inch (9.5 mm)
18L	7/16 inch (11 mm)
20L	1/2 inch (12 mm)
22L	9/16 inch (14 mm)
24 L	5/8 inch (15 mm)
27 L	10/16 inch (17 mm)
28L	11/16 inch (18 mm)
30L	3/4 inch (19 mm)
32L	13/16 inch (20 mm)
36L	7/8 inch (22 mm)
40L	1 inch (25 mm)
44L	1 1/8 inches (27 mm)
54L	1 3/8 inches (34 mm)
72L	1 3/4 inches (44 mm)

Note: Ligne is the unit of measurement that is used to indicate the diameter of a button. 1 ligne = 0.025 inch (0.635 mm).
Sources: http://www.mjtrim.com/pdf/buttons.pdf; http://www.naturalbuttons.com/Buttons%20Ligne.htm.

Lab Activity 3.5 Continued: Wear-Testing Comparison Project Garment

Go back to the worksheet for Lab Activity 3.5: Wear-Testing Comparison Project Garment. Update your wear log and refurbishment method for week 1. Measure the garment according to the specified method used, when you originally measured the garment. Calculate dimensional change for week 1, and record the results on worksheet 3.5. After you have completed this activity, move to Lab Activity 4.3.

Lab Activity 4.3: Comparison Project Garment Evaluation of Appearance and Color Change after Laundering

Review Chapter 12 of *Apparel Quality: A Guide to Evaluating Sewn Products* prior to the lab to fully understand the concepts and terminology used in this activity. For Lab Activity 4.3: Comparison Project Garment Evaluation of Appearance and Color Change after Laundering, each group member will bring his or her Comparison Project garments (wear-test and control garments) to lab for evaluation. This lab activity will be conducted in your Comparison Project lab groups (teams of four to five students). Each team member will compare his or her control garment to the wear-tested garment to determine if changes in color and appearance have occurred or if the garment has remained the same. Document your individual findings on worksheet 4.3.

Lab Activity 4.1: GARMENT ANALYSIS OF CONSTRUCTION DETAILS ACROSS PRICE CLASSIFICATIONS

Each team member, in collaboration with the rest of the team, will complete this worksheet for his or her selected garment. Document how each of the construction details on this worksheet are used to create the fit and style of the garment. Refer to Table 4.1 and the figures from Chapter 4 of the text for assistance with identification and use of proper terminology.

Your Name

Name(s) of Teammate(s)

Product Description	**Price Classification**

Fit and Construction Details

Ease
Does the garment possess the appropriate amount of functional/wearing ease to allow for adequate movement of the body? ☐Yes ☐No If no, explain which portions of the garment need more functional ease added.

Lab Activity 4.1 *(cont.)*

Fabric Grain
Which type(s) of grain is used in your garment? Explain where each type is used within the garment? If cross-grain or bias grain is used, explain the benefit it provides. ☐Straight grain ☐Cross-grain ☐Bias grain
Is the fabric cut on-grain or off-grain? ☐On-grain ☐Off-grain If the fabric is off-grain, was this noticeable at the time it was purchased? ☐Yes ☐No Has this impacted the comfort of the garment? Explain how.

Balance
Is the garment symmetric? ☐Yes ☐No Explain your answer.
If the garment is symmetric, does it hang evenly (extend from the body or contour the body) on both sides of the garment? ☐Yes ☐No

Lab Activity 4.1 *(cont.)*

Shaping Methods (Line)
Are darts present in your garment? ☐Yes ☐No If so, select each type and list garment location.
☐Straight ☐French ☐Concave ☐Convex ☐Double-ended
Are dart variations used in your garment? ☐Yes ☐No If so, select each type and list garment location.
☐Dart slash ☐Dart tucks ☐Flange darts

Lab Activity 4.1 *(cont.)*

Select the types of seams and dart equivalents present in your garment. List the garment locations for each type.
☐Straight seams ☐Princess line seams ☐Side panels ☐Gores ☐Yoke ☐Split yoke ☐Godet ☐Gussets
Are gathers or ruching present in your garment? ☐Yes ☐No If so, select the type and list garment location.
☐Gathers ☐Ruching
Are pleats present in your garment? ☐Yes ☐No If so, select the type and list garment location.
☐Accordion pleats ☐Knife pleats ☐Box pleats ☐Inverted pleats

Lab Activity 4.1 *(cont.)*

Are tucks present in your garment? ☐Yes ☐No If so, select the type and garment location.
☐Pin tucks ☐Spaced tucks ☐Blind tucks ☐Scalloped tucks ☐Cross tucks ☐Release tucks
Set
Are drag lines/wrinkles present anywhere on the garment caused by tension or slackness of the fabric? ☐Yes ☐No If yes, indicate each area of the garment that contains drag lines, and indicate the direction of the drag line (horizontal, vertical, or diagonal). *Horizontal drag lines* indicate the area of the garment is too tight to accommodate the curves of the body. *Vertical drag lines* indicate the area of the garment is too loose on the body. *Diagonal drag lines* indicate more shaping is needed in the particular area of the garment because it is too loose or too tight to accommodate the contours of the body.

Garment Openings and Closures		
Are facings present in your garment? ☐Yes ☐No If so, select the types and list garment locations.		
☐Shaped facing	☐Extended facing	
☐Bias facing		
☐Separate facing	☐Combination facing	
Are plackets present in your garment? ☐Yes ☐No If so, select the type and list garment location.		
☐Band placket	☐Bound placket	
☐Continuous lap placket		
☐Faced-slashed	☐Hemmed edge placket	
☐Tailored placket	☐Tailored placket with gauntlet button	
Does your garment have a closure(s)? ☐Yes ☐No If so, select the type and list garment location.		
☐Sew-through buttons	☐Shank buttons	☐Tack buttons
☐Straight buttonhole	☐Bound buttonhole	☐Slot buttonhole ☐Button loops
_______ Button ligne (See Table 4.1.)		
☐Center zipper insertion ☐Lap zipper	☐Exposed zipper insertion ☐Fly-front concealed zipper application	☐Invisible zipper insertion ☐Trouser fly zipper insertion
☐Sew-in snaps	☐No-sew snaps	
☐Hook and eyes	☐Hook and bar	
☐Ties		

Lab Activity 4.1 *(cont.)*

Does your garment have a waistband? ☐Yes ☐No If so, select the type and list garment location.		
☐Straight waistband	☐Contoured waistband	☐Bias-faced waistband
☐Trouser waistband	☐Curtain waistband	
☐Elasticized waistband with casing	☐Elasticized waistband without casing	
Hem Finishes		
What type of hem finish is present in the garment? Select the type and list garment location.		
☐Single-folded hem	☐Double-folded hem	☐Overedge hem
☐Rolled hem	☐Band hem	
☐Welt finished hem	☐Faced hem	

Answer the following questions about your team's findings after evaluating all of the garments across price categories.

Are there differences in the quality of the design elements and shaping methods used in the garments? **☐Yes ☐No** **Do the price classifications for each garment reflect these quality differences?** **☐Yes ☐No Explain your answers.**

Lab Activity 4.1 *(cont.)*

Are there differences in the types of findings used in each of the garments? ☐Yes ☐No Explain your answer.

Were the findings appropriate for the type of garment? ☐Yes ☐No Explain your answer.

Lab Activity 4.1 *(cont.)*

Compare the quality of the findings between price classifications.

Were there any construction details you expected to see in a garment from a particular price category? **☐Yes ☐No Explain your answer.**

Lab Activity 4.1 *(cont.)*

Were there any construction details you were surprised to see in a garment from a particular price category? ☐Yes ☐No Explain your answer.

Lab Activity 4.2: COMPARISON PROJECT GARMENT ANALYSIS OF CONSTRUCTION DETAILS ACROSS COMPETING BRANDS

Each team member, in collaboration with the test of the team, will complete this worksheet for his or her selected garment. Document how each of the construction details on this worksheet are used to create the fit and style of the garment. Refer to Table 4.1 and the figures from Chapter 4 of the text for assistance with identification and use of proper terminology.

Your Name

Name(s) of Teammate(s)

Product Description—describe the style of the garment	Price Classification

Fit and Construction Details

Ease
Does the garment possess the appropriate amount of functional/wearing ease to allow for adequate movement of the body? ☐Yes ☐No If no, explain which portions of the garment need more functional ease added.

Lab Activity 4.2 *(cont.)*

Fabric Grain
Which type(s) of grain is used in your garment? Explain where each type is used within the garment? If cross-grain or bias grain is used explain the benefit it provides.
☐Straight grain ☐Bias grain
Is the fabric cut on-grain or off-grain? ☐On-grain ☐Off-grain If the fabric is off-grain, was this noticeable at the time it was purchased? ☐Yes ☐No Has this impacted the comfort of the garment? Explain how.

Balance
Is the garment symmetric? ☐Yes ☐No Explain your answer.
If the garment is symmetric, does it hang evenly (extend from the body or contour the body) on both sides of the garment? ☐Yes ☐No

Lab Activity 4.2 *(cont.)*

Shaping Methods (Line)
Are straight darts present in your garment? ☐Yes ☐No If so, list garment location.
Select the types of seams and dart equivalents present in your garment? List the garment locations for each type.
☐Straight seams ☐Shaped seams
Are gathers or ruching present in your garment? ☐Yes ☐No If so, select the type and list garment location.
☐Gathers ☐Ruching
Set
Are drag lines/wrinkles present anywhere on the garment caused by tension or slackness of the fabric? ☐Yes ☐No If yes, indicate each area of the garment that contains drag lines, and indicate the direction of the drag line (horizontal, vertical, or diagonal). *Horizontal drag lines* indicate the area of the garment is too tight to accommodate the curves of the body. *Vertical drag lines* indicate the area of the garment is too loose on the body. *Diagonal drag lines* indicate more shaping is needed in the particular area of the garment because it is too loose or too tight to accommodate the contours of the body.

Lab Activity 4.2 *(cont.)*

Garment Openings and Closures
Are plackets present in your garment? ☐Yes ☐No If so, select the type and list garment location.
☐Band placket ☐Bound placket ☐Continuous lap placket
Does your garment have a closure(s)? ☐Yes ☐No If so, select the type and list garment location.
☐Sew-through buttons ☐Straight buttonhole ______ Button ligne (See Table 4.1.) ☐Sew-in snaps ☐No-sew snaps
Hem Finishes
What type of hem finish is present in the garment? Select the type and list garment location.
☐Single-folded hem ☐Double-folded hem ☐Overedge hem

Lab Activity 4.2 *(cont.)*

Answer the following questions about your team's findings after evaluating all of the garments across competing brands.

Are there any construction details you would have expected to see in a garment from this price category that are not present? ☐Yes ☐No Explain your answer.

Are there any construction details you were surprised to see in one or more of the competing brands? ☐Yes ☐No Explain your answer.

Lab Activity 4.2 *(cont.)*

Do the construction details for each competing brand correspond with your group's expectations for this price category? ☐Yes ☐No Explain your answer.

Lab Activity 4.3: COMPARISON PROJECT GARMENT EVALUATION OF APPEARANCE AND COLOR CHANGE AFTER LAUNDERING

Each team member will complete a worksheet for his or her selected garment.

Your Name

Name(s) of Teammate(s)

Product Description—describe the style of the garment	**Price Classification**

Test Method	**Name of Test**	**Report**
ASTM 6321	Standard Practice for the Evaluation of Machine Washable T-Shirts Section 7.4.4 Shade Difference	Report color change rating
AATCC Evaluation Procedure 1	Gray Scale for Color Change	Report color change rating
Color Permanence to Home Laundering	**Color Change Rating** (also indicate if the worn shirt has lost or gained color)	**Note Changes in Appearance** (i.e., surface abrasion, puckering of seams, roping of hems, snagging, hand, holes, bagging)
Week 1		
Week 2		
Week 3		
Week 4		
Week 5		

NOTES

CHAPTER 5

Apparel Sizing and Fit Strategies Lab

LAB OBJECTIVES:

- To gather body measurement data for determining apparel sizing.
- To examine garments to evaluate sizing and fit.

In Chapter 5 of *Apparel Quality: A Guide to Evaluating Sewn Products*, the importance of body measurement data and the methods used for gathering this information are discussed to further the understanding of how size standards for apparel are developed. Domestic and international voluntary size standards provide retailers with sizing information that reflects the body shapes and sizes of today's consumers. Many brands have established their own sizing systems but can utilize these standards for updating their sizes. Prototype development and evaluating garment fit are important aspects in the design of apparel products that can lead to customer satisfaction or returned merchandise. Technical packages and specifications for apparel products also help ensure the desired quality level and leave little for interpretation during production.

There is one lab activity (worksheet 5.1) for Chapter 5 and three additional activities carried over from Chapters 3 and 4 (preparation of the piece goods specimen for Lab Activity 3.2 and the Comparison Project garment analysis after week 2 of the wear test for Lab Activities 3.5 and 4.3). Chapter 5 lab activities will focus on analyzing body measurements, fit, and garment sizing. The elements of fit (ease, grain, line, balance, and set) were evaluated as part of Lab Activities 4.1 and 4.2. Any work that has not been completed for the Chapter 3 lab activities, such as cutting and preparing the piece goods specimens, should be completed during this lab session so they are ready for testing in a future lab session.

PREPARATION AND SUPPLIES FOR CHAPTER 5 LAB ACTIVITIES

Prior to Chapter 5 lab activities, wear your comparison project garment for the designated number of hours and clean it according to the garment label care instructions. Both the wear test garment and the control must be brought to lab for evaluation. Bring a tape measure for taking body measurements and dress accordingly for Chapter 5 lab activities so that accurate body measurements can be taken. If your team has not completed preparation of templates, the sampling plan and marking and cutting of piece goods specimens the following supplies from Chapter 3 lab activities will be needed: 4 pages of card stock or 1 package of Visi-GRID Quilter's Template Sheets (contains

4 plastic grid sheets, 8 ½ inches by11inches or metric equivalent), graph paper, a permanent marking pen (such as Sharpie rub-a-dub), ruler, scissors, 2 additional pieces of card stock for mounting specimens, and 2 yard/2 meter lab sample of piece goods.

Lab Activity 5.1: Comparison Project Garment Analysis of Size and Fit

For Lab Activity 5.1: Comparison Project Garment Analysis of Size and Fit, work in your Comparison Project teams of four to five people. Each group member will use his or her Comparison Project control garment for evaluation. A tape measure is needed to complete this activity. As a team, begin completing worksheet 5.1. Consult Table 5.1 for men's and misses' body measurements by size range. After you have completed worksheet 5.1, move on to Lab Activity 3.5.

Table 5.1 Misses' and Men's Body Measurements by Size Range

MISSES' BODY MEASUREMENTS AND CORRESPONDING SIZE RANGES						
SIZE	00 / XS	0 / XS	2 / S	4 / S	6 / M	8 / M
BODY MEASUREMENTS IN INCHES (CENTIMETERS)						
Bust girth	31 1/8 (79.06)	31 3/4 (80.65)	33 (83.82)	34 1/8 (86.68)	35 1/2 (89.54)	36 1/4 (92.08)
Waist girth—curvy	29 7/8 (60.64)	24 5/8 (62.55)	25 3/8 (64.45)	26 1/8 (66.36)	27 (68.58)	28 (71.12)
Waist girth—straight	25 3/8 (64.45)	26 1/8 (66.36)	26 7/8 (68.26)	27 5/8 (70.17)	28 1/2 (72.39)	29 1/2 (74.93)
Armscye girth	14 1/2 (36.83)	14 3/4 (37.47)	15 1/8 (38.42)	15 1/2 (39.37)	15 3/4 (40.01)	16 (40.64)
Upper-arm girth	9 3/4 (24.77)	10 (25.40)	10 1/4 (26.04)	10 1/2 (26.67)	10 3/4 (27.31)	11 1/8 (28.26)
SIZE	10 / L	12 / L	14 / L	16 / XL	18 / XL	20 / XL
BODY MEASUREMENTS IN INCHES (CENTIMETERS)						
Bust girth	37 1/4 (94.62)	38 3/4 (98.43)	40 3/8 (102.55)	42 1/8 (107.00)	44 (111.76)	46 (116.84)
Waist girth—curvy	29 (73.66)	30 3/4 (78.11)	32 1/2 (82.55)	34 1/2 (87.63)	36 3/4 (93.35)	39 (99.06)
Waist girth—straight	30 1/2 (77.47)	32 1/4 (81.92)	34 (86.36)	36 (91.44)	38 1/4 (97.16)	40 1/2 (102.87)
Armscye girth	16 1/4 (41.28)	16 7/8 (42.86)	17 1/2 (44.45)	18 1/8 (46.04)	18 7/8 (47.94)	19 5/8 (49.85)
Upper-arm girth	11 1/2 (29.21)	11 3/4 (29.85)	12 1/8 (30.80)	12 1/2 (31.75)	13 (33.02)	13 5/8 (34.61)

Table 5.1 *(cont.)*

MEN'S BODY MEASUREMENTS AND CORRESPONDING SIZE RANGES						
SIZE	**34 / S**	**35 / S**	**36 / S**	**37 / S**	**38 / M**	**39 / M**
BODY MEASUREMENTS IN INCHES (CENTIMETERS)						
Chest girth	34 (86.36)	35 (88.90)	36 (91.44)	37 (93.98)	38 (96.52)	39 (99.06)
Waist girth	28 1/2 (72.39)	29 1/2 (74.93)	30 1/2 (77.47)	31 1/2 (80.01)	32 1/2 (82.55)	33 1/2 (85.09)
Armscye girth—short	15 1/4 (38.74)	15 1/2 (39.37)	15 3/4 (40.01)	16 1/8 (40.96)	16 1/2 (41.91)	16 7/8 (42.86)
Armscye girth—*regular*	15 5/8 (39.69)	15 7/8 (40.32)	16 1/8 (40.96)	16 1/2 (41.91)	16 7/8 (42.86)	17 1/4 (43.82)
Armscye girth—*tall*	16 (40.64)	16 1/4 (41.28)	16 1/2 (41.91)	16 7/8 (42.86)	17 1/4 (43.82)	17 5/8 (44.77)
Upper-arm girth	11 3/4 (29.85)	12 (30.48)	12 1/4 (31.12)	12 1/2 (31.75)	12 3/4 (32.39)	13 (33.02)
SIZE	**40 / M**	**41 / M**	**42 / L**	**43 / L**	**44 / L**	**45 / L**
BODY MEASUREMENTS IN INCHES (CENTIMETERS)						
Chest girth	40 (101.60)	41 (104.14)	42 (106.68)	43 (109.22)	44 (111.76)	45 (114.30)
Waist girth	34 1/2 (87.63)	35 1/2 (90.17)	36 1/2 (92.71)	37 1/2 (95.25)	38 1/2 (97.79)	39 3/4 (100.97)
Armscye girth—short	17 1/2 (43.82)	17 1/2 (44.45)	17 3/4 (45.09)	18 (45.72)	18 3/8 (46.67)	18/58 (47.31)
Armscye girth—regular	17 5/8 (44.77)	17 7/8 (45.40)	18 1/8 (46.04)	18 3/8 (46.67)	18 3/4 (47.63)	19 (48.26)
Armscye girth—tall	18 (45.72)	18 1/4 (46.36)	18 1/2 (46.99)	18 3/4 (47.63)	19 1/8 (48.58)	19 3/8 (49.21)
Upper-arm girth	13 1/4 (33.66)	13 1/2 (34.29)	13 3/4 (34.93)	14 (35.56)	14 1/4 (36.20)	14 1/2 (36.83)

Note: ASTM does not have a standard for men's sizing between boys size 20 and mature men's size 34.

Sources: ASTM International, "D 5585 Standard Tables of Body Measurements for Adult Female Misses Figure Type, Size Range 00–20," West Conshohoken, PA: ASTM International, 2014; ASTM International "D6240 Standard Tables of Body Measurements for Mature Men Ages 35 and Older, Sizes Thirty-Four to Fifty-Two (34 to 52) Short, Regular, and Tall," West Conshohoken, PA: ASTM International, 2014.

Lab Activity 3.5 Continued: Wear-Testing Comparison Project Garment

Go back to the worksheet for Lab Activity 3.5: Wear-Testing Comparison Project Garment. Update your wear log and refurbishment method for week 2. Measure the garment according to the specified method used when you originally measured the garment. Calculate the dimensional change for week 2 ,and record the results on worksheet 3.5. After you have completed this activity, proceed to Lab Activity 4.3.

Lab Activity 4.3 Continued: Comparison Project Garment Evaluation of Appearance and Color Change after Laundering

For Lab Activity 4.3, each group member will bring his or her Comparison Project garments (wear-test and control garments) to lab for evaluation. Each team member will compare his or her control garment to the wear-tested garment for week 2 to determine if changes in color and appearance have occurred or if the garment has remained the same. Document your individual findings for week 2 on worksheet 4.3. After you have updated the worksheet, finish cutting, marking, and preparing specimens from the piece goods lab sample so they are ready for testing in a future lab session.

Lab Activity 5.1: COMPARISON PROJECT GARMENT ANALYSIS OF SIZE AND FIT

Each team member will evaluate and complete individual worksheets for his or her control garment. A tape measure is needed to complete this activity. Refer to Table 5.1 to assist with body measurements and size.

Your Name

Name(s) of Teammate(s)

Product Description—describe the style of the garment	Price Classification

Garment Fit Type

Select the fit type that best represents the fit of your Comparison Project garment.

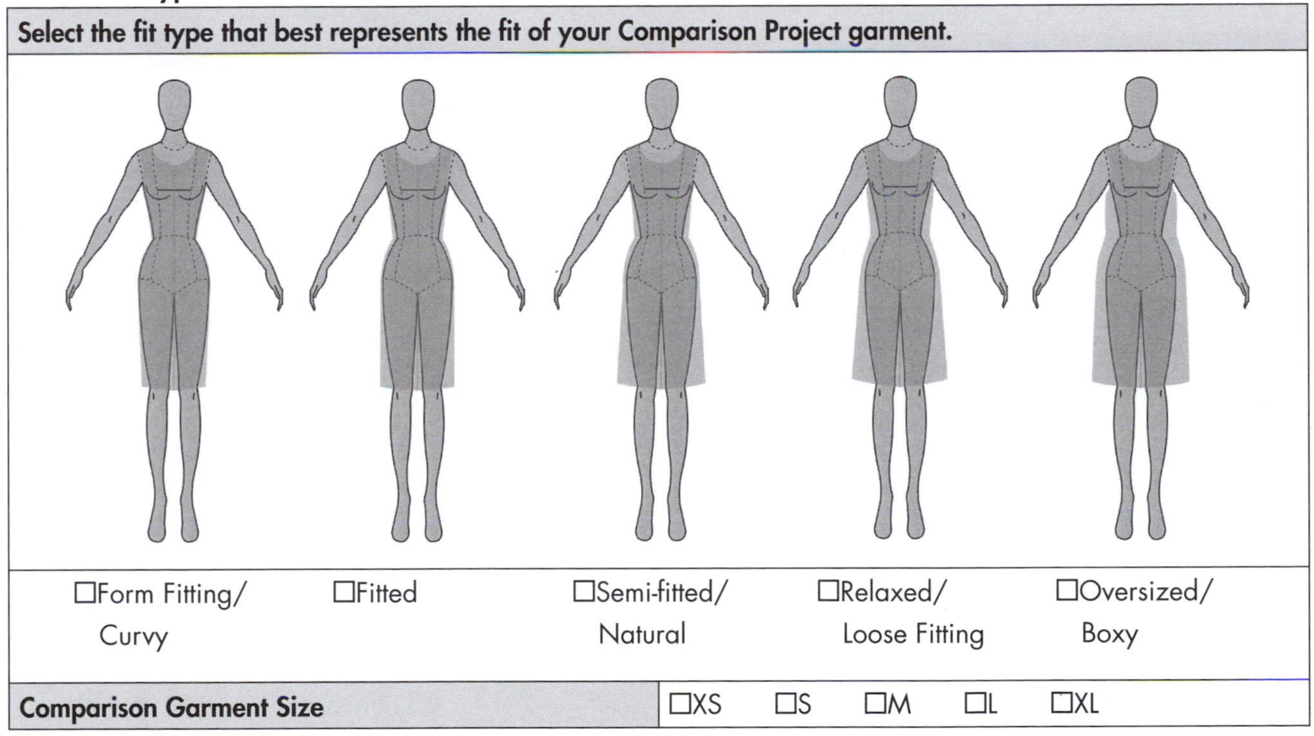

☐Form Fitting/ Curvy	☐Fitted	☐Semi-fitted/ Natural	☐Relaxed/ Loose Fitting	☐Oversized/ Boxy

Comparison Garment Size	☐XS ☐S ☐M ☐L ☐XL

Lab Activity 5.1 *(cont.)*

Measure your body and record the measurement below. Use Table 5.1 and record the ASTM size designation for each measurement.

Garment Measurements and Size

Body Measurement	Your Body Measurements	ASTM Size Designation (XS–XL)
Bust/chest girth Measure the circumference across the fullest part of the chest with arms relaxed at sides.		
Waist girth Measure the circumference of the smallest part of the waistline.		☐Misses curvy ☐Misses straight ☐Men's
Armscye girth Measure the circumference starting at the top of the shoulder joint, under the arm and back around to the starting point. Take measurement with arms relaxed at sides.		☐Misses ☐Mens short ☐Mens regular ☐Mens tall
Upper-arm girth Measure the circumference in the middle of the arm between the shoulder joint and the elbow.		
Do all of your measurements fit within one size range (XS–XL)? ☐Yes ☐No If no, select the area(s) that fell into another size range.		
☐Bust/Chest Girth ☐Waist Girth ☐Armscye ☐Upper-Arm Girth		

Lab Activity 5.1 *(cont.)*

Do you think the measurement difference is due to the amount of functional ease or design ease added by the brand? ☐Functional Ease ☐Design Ease Explain your answer.

Does the size designated for your shirt brand correspond to your ASTM size designation? ☐Yes ☐No If no, how does it differ?

When comparing your results with your teammates, are there significant differences between the brands in relation to body measurements and ASTM sizing? ☐Yes ☐No If yes, how do they differ?

NOTES

CHAPTER 6

ASTM and ISO Stitch and Seam Classifications Lab

LAB OBJECTIVES:

- To examine a garment and identify the ASTM International (ASTM) or International Organization for Standardization (ISO) stitch and seam classifications and to determine the specific stitches and seams used.
- To determine the number of stitches per inch or stitches per centimeter for designated areas of a garment.
- To calculate thread consumption for a garment.

Chapters 6 and 7 of *Apparel Quality: A Guide to Evaluating Sewn Products* focus on standards for designating stitches and seams used in the apparel industry for garment assembly. Knowledge of thread and fabric construction, stitch classifications, and seam efficiency in relation to the garments end use helps designers and product developers make choices appropriate for the assembly of apparel items.

There are two lab activities (worksheets 6.1, 6.2) for Chapters 6, and two additional activities carried over from Chapters 3 and 4 (the Comparison Project garment analysis after week 3 of the wear test for Lab Activities 3.5 and 4.3). The lab activities for Chapter 6 will focus on identifying the major stitch and seam classifications used in your comparison garment. You will use the comparison shirt designated for physical testing and deconstruction and analysis of stitches and seams. Do not use the wear-test or control garments. Deconstruct the stitches and seams to determine each specific stitch and seam contained in the garment. In addition, you will use the ANECALC to calculate the thread consumption for your comparison shirt.

PREPARATION AND SUPPLIES FOR CHAPTER 6 LAB ACTIVITIES

Prior to Chapter 6 lab activities, wear your comparison project garment for the designated number of hours and clean it according to the garment label care instructions. All of the comparison project garments (wear test garment, control, and physical testing garment) must be brought to lab for evaluation. Each team member (groups of 4 to 5) will bring a seam ripper or dissecting pick, ruler, and fabric scissors.

Lab Activity 6.1: Comparison Project Garment Identification of Stitches and Seams

For Lab Activity 6.1: Comparison Project Garment Identification of Stitches and Seams, work in Comparison Project teams of four to five people. Each group member will use his or her Comparison Project garment designated for physical testing for evaluation. A seam ripper or dissecting pick, a ruler, and a pair of fabric scissors are needed for this lab activity. Each team member will complete worksheet 6.1. Refer to Chapter 6 and 7 of *Apparel Quality: A Guide to Evaluating Sewn Products* as well as to the following standards for assistance with identifying each of the stitches and seams contained in your Comparison Project control garment: ASTM D6193 Standard Practice for Stitches and Seams, ISO 4915 Textiles—Stitch Types—Classification and Terminology, and ISO 4916 Textiles—Seam Types—Classification and Terminology. As you deconstruct portions of the garment, take care to only snip a small portion of the seams and stitching because the remainder of the shirt will be used for cutting additional specimens for physical testing in future lab sessions. If your school does not have access to these standards, go ahead and identify each of the stitches and seams using their main classifications outlined in the text rather than deconstructing the garment to identify the exact stitches and seams used.

Lab Activity 6.2: Comparison Project Garment Thread Consumption Calculations

You may need to complete this activity outside of class because it requires the use of a computer. Log onto http://www.amefird.com/technical-tools/thread-consumption/anecalc/ or go to http://www.amefird.com, click on the Technical Tools drop down menu and select Thread Consumption and then ANECALC to open the ANECALC Apparel Guidelines. Find the American & Efird Thread Consumption Calculator for the type of shirt that represents your comparison garment. Using Table 6.1, find the put-up price per cone and color for each of the thread types used in this garment and enter the pricing next to the respective thread type at the bottom, middle of the page. Change the stitches to those represented in your shirt. The following information will be provided: thread price per yard, thread price per garment, number of cones of thread needed for each thread type, needle thread yards, bobbin thread yards, looper thread yards, total yards per operation, total yards per garment consumed, yards per garment of waste, total yards per garment including waste, and cost per garment. Print a copy of your thread consumption calculation to turn in.

Table 6.1 Thread Pricing for American and Efird Thread Consumption Calculator

THREAD TYPE	YARDS PER CONE	PUT-UP PRICE PER CONE—WHITE THREAD	PUT-UP PRICE PER CONE—COLORED THREAD
T-18 Perma core	6,000 yd	$4.88	$5.93
T-21 Perma spun	6,000 yd	$3.37	$3.74
T-24 Perma core	6,000 yd	$5.57	$6.08
T-27 Perma spun	6,000 yd	$3.42	$3.94
T-40 Perma core	6,000 yd	$7.06	$7.79
T-18 Wildcat plus	26,800 yd in white	$6.64	
	25,423 yd in colors		$12.69
T-24 Wildcat plus	20,247 yd in white	$6.64	
	18,355 yd in colors		$12.69

Sources: American & Efird, 2014.

Lab Activity 3.5 Continued: Wear-Testing Comparison Project Garment

Go back to worksheet 3.5. Update your wear log and refurbishment method for week 3. Measure the garment according to the specified method used when you originally measured the garment. Calculate dimensional change for week 3, and record the results on worksheet 3.5. After you have completed this activity, move on to Lab Activity 4.3: Comparison Project Garment Evaluation of Appearance and Color Change after Laundering.

Lab Activity 4.3 Continued: Comparison Project Garment Evaluation of Appearance and Color Change after Laundering

For Lab Activity 4.3, each group member will bring his or her Comparison Project garments to lab for evaluation. Each team member will compare his or her control garment to the wear-tested garment for week 3 to determine if changes in color and appearance have occurred or if the garment has remained the same. Document your individual findings for week 3 on worksheet 4.3.

Lab Activity 6.1: COMPARISON PROJECT GARMENT IDENTIFICATION OF STITCHES AND SEAMS

Each member will evaluate and complete this worksheet for his or her comparison garment. A seam ripper or dissecting pick, ruler, and scissors are needed for this activity.

Your Name

Name(s) of Teammate(s)

Product Description—describe the style of the garment	Price Classification

Stitches and Seams

ASTM D6193 Standard Practice for Stiches and Seams		ISO 4915 Textiles—Stitch Types—Classification and Terminology		ISO 4916 Textiles—Seam Types—Classification and Terminology	
Garment Location	**Stitch Classification**	**Stitch Number**	**SPI/SPC**	**Seam Classification**	**Seam Number**
Neckline					
Taped back neckline					
Front placket					
Shoulder seam					
Sleeve seam					
Sleeve hem					

Lab Activity 6.1 *(cont.)*

Garment Location	Stitch Classification	Stitch Number	SPI/SPC	Seam Classification	Seam Number
Side seam					
Bottom hem					
Pocket attachment					
Pocket top edge finish					

How many stitch classifications were found in your garment? How many different stitches?	_______ Stitch Classifications _______ Stitches
How many seam classifications were found in your garment? How many different seams?	_______ Seam Classifications _______ Seams

Lab Activity 6.1 *(cont.)*

Discuss and compare your findings with your teammates.

Was there a difference in the stitches and SPI/SPC used for each of the brands? ☐Yes ☐No If yes, describe the differences.

Was there a difference in the seams used for each of the brands? ☐Yes ☐No If yes, describe the differences.

Lab Activity 6.1 *(cont.)*

Which brand(s) added the most value to the product based on the seams used? Explain your answer.

NOTES

CHAPTER 7

Labeling Regulations and Guidelines for Manufacturing Apparel Lab

LAB OBJECTIVES:

- To examine garment labels for compliance with government regulations.

Chapter 9 of *Apparel Quality: A Guide to Evaluating Sewn Products* focuses on garment-labeling regulations for the major consumer markets and the similarities and differences between each country's mandatory and voluntary regulations. In today's global marketplace, manufacturers are selling apparel to different countries around the world and are choosing to integrate the highest regulatory standards into their products to ensure that individual mandatory regulations for labeling are met. Occasionally, there are garments that enter into commerce that are not 100 percent compliant with federal labeling regulations.

There are two lab activities (worksheets 7.1, 7.2) for this chapter and two additional activities carried over from Chapters 3 and 4 (Lab Activities 3.5 and 4.3 Comparison Project garment analysis after week 4 of the wear test). The lab activities for this chapter will focus on evaluating the content of garment labels to determine if they comply with government regulations. Care methods will also be evaluated.

PREPARATION AND SUPPLIES FOR CHAPTER 7 LAB ACTIVITIES

Prior to Chapter 7 lab activities, wear your comparison project garment for the designated number of hours and clean it according to the garment label care instructions. Both the wear test garment and the control must be brought to lab for evaluation. In preparation for Chapter 7 lab activities meet with your small group (2 to 3 people). Each group member will bring a different type of garment from their wardrobe (i.e. a pair of jeans, a sweater, a shirt). Determine which garment each person will bring to avoid duplication.

Lab Activity 7.1: Comparison Project Garment Label Compliance

For Lab Activity 7.1: Comparison Project Garment Label Compliance, work in your Comparison Project teams of four to five people, and each group member will use his or her Comparison Project control garment for evaluation. Begin by reading the labels in your Comparison Project garment.

Lab Activity 7.2: Garment Care

For Lab Activity 7.2: Garment Care, work in teams of two to three people. Each group member will bring a different type of garment from his or her wardrobe (i.e., a pair of jeans, a sweater, a shirt). Begin by reading the care label information for each of the garments. Evaluate the recommended care methods used, and determine if the information is complete and in compliance with federal regulations. See Figures 7.1, 7.2, and 7.3 for ASTM, ISO/GINETEX, and JIS care symbols.

Lab Activity 3.5 Continued: Wear-Testing Comparison Project Garment

Go back to Lab Activity 3.5: Wear-Testing Comparison Project Garment worksheet. Update your wear log and refurbishment method for week 4. Measure the garment according to the specified method used, when you originally measured the garment. Calculate dimensional change for week 4, and record the results on worksheet 3.5. After you have completed this task, proceed to Lab Activity 4.3: Comparison Project Garment Evaluation of Appearance and Color Change after Laundering.

Lab Activity 4.3 Continued: Comparison Project Garment Evaluation of Appearance and Color Change after Laundering

For Lab Activity 4.3, each group member will bring his or her Comparison Project garments to lab for evaluation. Each team member will compare his or her control garment to the wear-tested garment for week 4 to determine if changes in color and appearance have occurred or if the garment has remained the same. Document your individual findings for week 4 on worksheet 4.3.

Figure 7.1 ASTM Guide to Care Symbols

Valued Quality. Delivered.

American Care Labeling

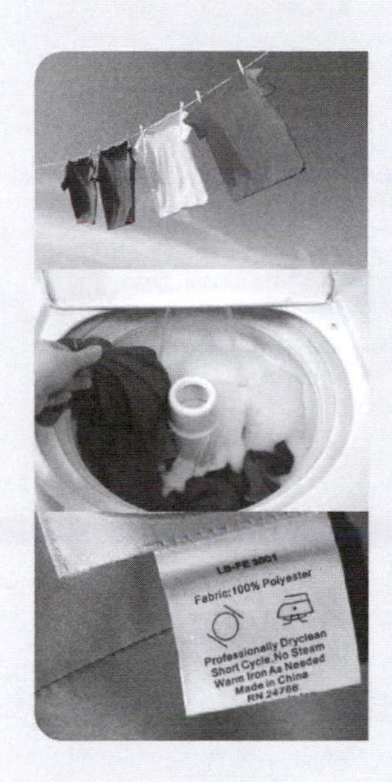

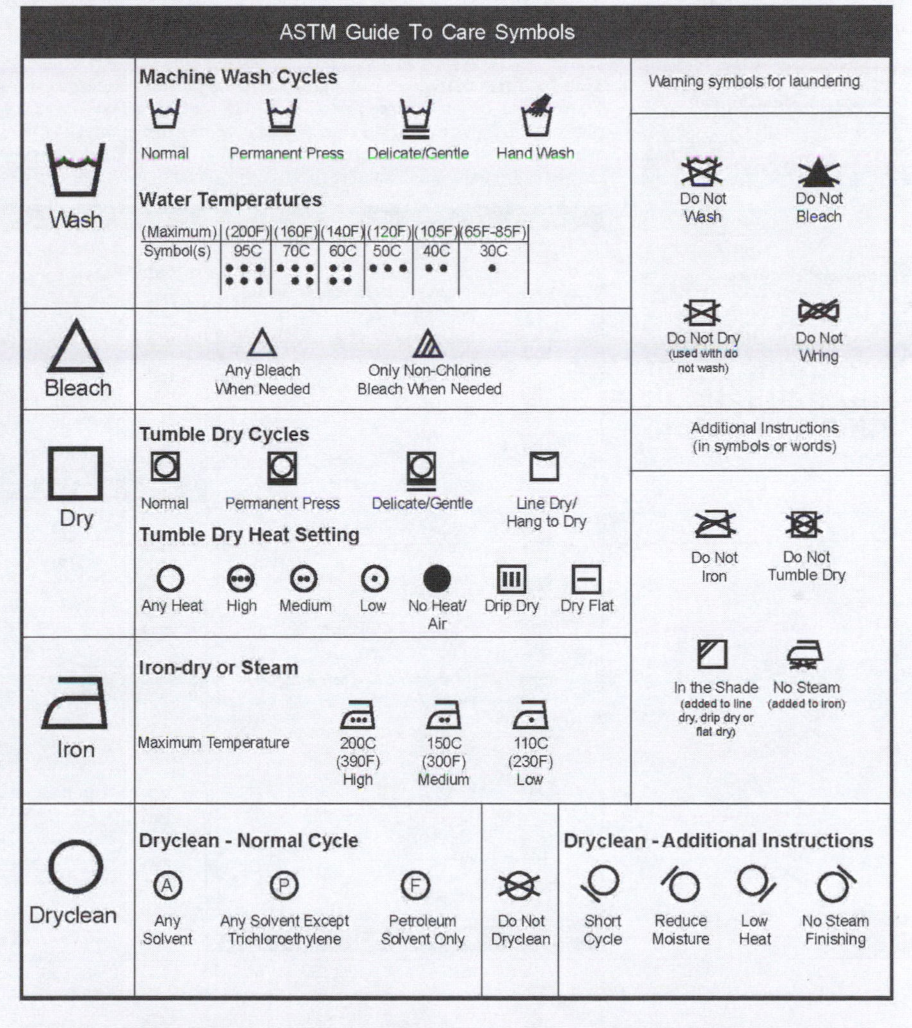

Note: This figure illustrates the symbols to use for laundering and drycleaning instructions. As a minimum, laundering instructions should include, in order, four symbols: washing, bleaching, drying and ironing; drycleaning instructions shall include one symbol. Additional words may be used to clarify the instructions.

I-CABD

Figure 7.2 ISO/GINETEX Care Symbols

Valued Quality. Delivered.

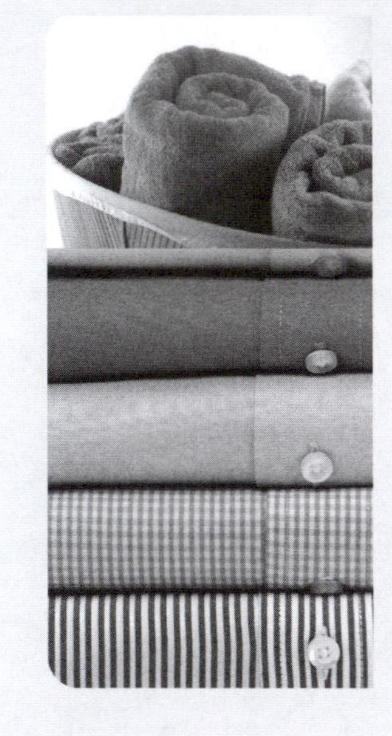

European Care Labeling

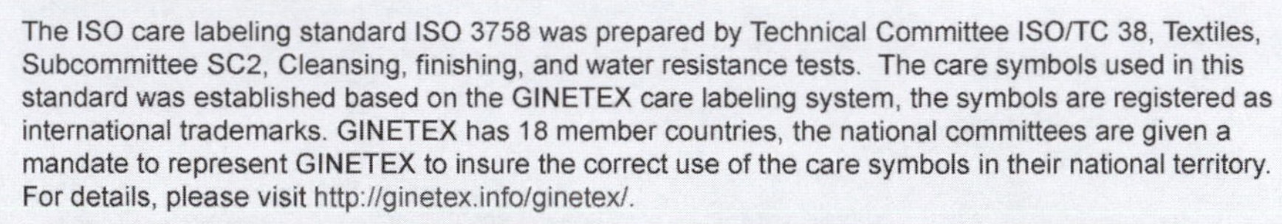
The ISO care labeling standard ISO 3758 was prepared by Technical Committee ISO/TC 38, Textiles, Subcommittee SC2, Cleansing, finishing, and water resistance tests. The care symbols used in this standard was established based on the GINETEX care labeling system, the symbols are registered as international trademarks. GINETEX has 18 member countries, the national committees are given a mandate to represent GINETEX to insure the correct use of the care symbols in their national territory. For details, please visit http://ginetex.info/ginetex/.

The first and second editions ISO 3758 were published in 1991 and 2005. The third edition ISO 3758:2012 has been published and replaces the previous version of the standard (ISO 3758:2005). Key changes are the addition of symbols for natural drying processes and the change of 'Do not bleach' symbol. The previous version used a blackened triangle; in the 2012 version this has now reverted back to a lined version. The care symbols used in ISO 3758: 2012 consist of 5 main treatments and shall appear in the order washing, bleaching, drying, ironing and professional textile care.

Symbol	Washing Process
95	- maximum washing temperature 95°C normal process
70	- maximum washing temperature 70°C normal process
60	- maximum washing temperature 60°C normal process
60	- maximum washing temperature 60°C mild process
50	- maximum washing temperature 50°C normal process
50	- maximum washing temperature 50°C mild process
40	- maximum washing temperature 40°C normal process
40	- maximum washing temperature 40°C mild process
40	- maximum washing temperature 40°C very mild process
30	- maximum washing temperature 30°C normal process
30	- maximum washing temperature 30°C mild process
30	- maximum washing temperature 30°C very mild process
	- wash by hand maximum temperature 40°C
	- do not wash

Symbol	Bleaching Process
	- any bleaching agent allowed
	- only oxygen / non-chlorine bleach allowed
	- do not bleach

Symbol	Tumble Drying Process
	- tumble drying possible - normal temperature - maximum exhaust temperatue 80°C
	- tumble drying possible - drying at lower temperature - maximum exhaust temperature 60°C
	- do not tumble dry

Symbol	Natural Drying Process
	- line drying
	- drip line drying
	- flat drying
	- drip flat drying
	- line drying in the shade
	- drip line drying in the shade
	- flat drying in the shade
	- drip flat drying in the shade

Symbol	Ironing Process
	- iron at a maximum sole-plate temperature of 200°C
	- iron at a maximum sole-plate temperature of 150°C
	- iron at a maximum sole-plate temperature of 110°C without steam steam ironing may cause irreversible damage
	- do not iron

Symbol	Professional Textile Care Process
P	- professional dry cleaning in tetrachloroethene and all solvents listed for the symbol F normal process
P	- professional dry cleaning in tetrachloroethene and all solvents listed for the symbol F mild process
F	- professional dry cleaning in hydrocarbons (distillation temperature between 150°C and 210°C, flash point between 38°C and 70°C) normal process
F	- professional dry cleaning in hydrocarbons (distillation temperature between 150°C and 210°C, flash point between 38°C and 70°C) mild process
	- do not dry clean
W	- professional wet cleaning normal process
W	- professional wet cleaning mild process
W	- professional wet cleaning very mild process
W	- do not professional wet clean

I-CABD

Figure 7.3 JIS Care Symbols

Valued Quality. Delivered.

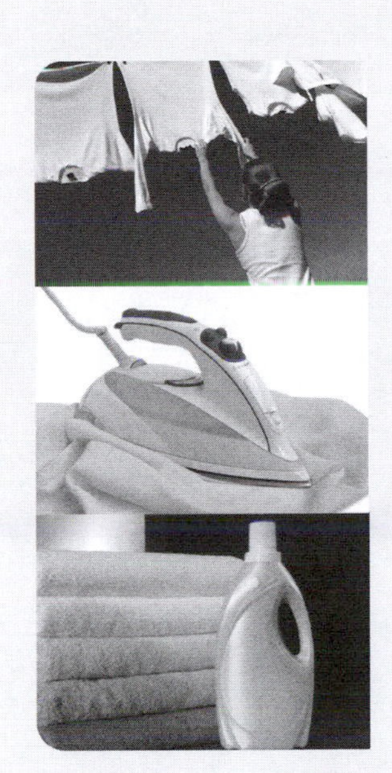

Japanese Care Labeling

Japanese care instructions, like other care label systems, must be in a specified order. This order is washing, bleaching, ironing, dry-cleaning, wringing, and drying.

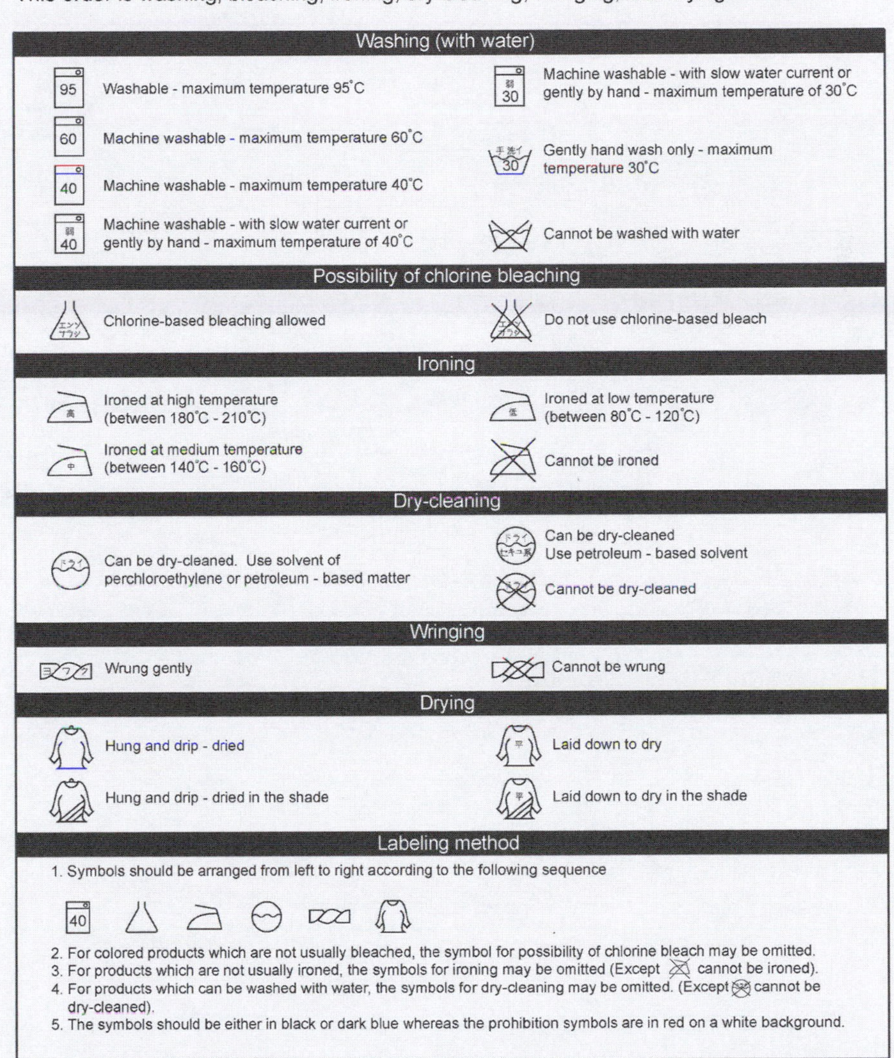

Washing (with water)

- 95 Washable - maximum temperature 95˚C
- 60 Machine washable - maximum temperature 60˚C
- 40 Machine washable - maximum temperature 40˚C
- 弱 40 Machine washable - with slow water current or gently by hand - maximum temperature of 40˚C
- 弱 30 Machine washable - with slow water current or gently by hand - maximum temperature of 30˚C
- 手洗イ 30 Gently hand wash only - maximum temperature 30˚C
- Cannot be washed with water

Possibility of chlorine bleaching

- エンソサラシ Chlorine-based bleaching allowed
- Do not use chlorine-based bleach

Ironing

- 高 Ironed at high temperature (between 180˚C - 210˚C)
- 中 Ironed at medium temperature (between 140˚C - 160˚C)
- 低 Ironed at low temperature (between 80˚C - 120˚C)
- Cannot be ironed

Dry-cleaning

- ドライ Can be dry-cleaned. Use solvent of perchloroethylene or petroleum - based matter
- ドライ セキユ系 Can be dry-cleaned Use petroleum - based solvent
- Cannot be dry-cleaned

Wringing

- ヨワク Wrung gently
- Cannot be wrung

Drying

- Hung and drip - dried
- Hung and drip - dried in the shade
- 平 Laid down to dry
- 平 Laid down to dry in the shade

Labeling method

1. Symbols should be arranged from left to right according to the following sequence
2. For colored products which are not usually bleached, the symbol for possibility of chlorine bleach may be omitted.
3. For products which are not usually ironed, the symbols for ironing may be omitted (Except [symbol] cannot be ironed).
4. For products which can be washed with water, the symbols for dry-cleaning may be omitted. (Except [symbol] cannot be dry-cleaned).
5. The symbols should be either in black or dark blue whereas the prohibition symbols are in red on a white background.

Attached remarks

1. The word '中性' indicates that neutral detergent should be used.

2. The symbol '' indicates that clothes ironed should be covered with another cloth.

3. The sentence 'ネット使用' indicates that a net should be used.

I-CABD

Lab Activity 7.1: COMPARISON PROJECT GARMENT LABEL COMPLIANCE

Each member will evaluate and complete this worksheet for his or her comparison garment.

Your Name

Name(s) of Teammate(s)

Product Description—describe the style of the garment	Price Classification

Write the information exactly the way it is listed on the garment label, and indicate whether the information is mandatory (M) or voluntary (V) and if the garment is in compliance for each of the following areas.

Labeling Content Evaluation

	Label Content	M or V	Compliant
Fiber content			☐Yes ☐No
Country of origin			☐Yes ☐No
Manufacturer or importer ID number			☐Yes ☐No
Care instructions			☐Yes ☐No
Size			☐Yes ☐No

Lab Activity 7.1 *(cont.)*

Is there any information that is missing? ☐Yes ☐No If yes, what is missing?

Lab Activity 7.2: GARMENT CARE

Team members will collaboratively evaluate the garments. Each individual will record the information pertaining to his or her garment on this worksheet. Teams of two will evaluate two garments; teams of three will evaluate three symbols.

Your Name

Name(s) of Teammate(s)

Product Description of Your Garment—describe the style of the garment

List the fiber content as it appears on the garment label.

What regulatory body oversees and enforces care labeling in apparel products?

Lab Activity 7.2 *(cont.)*

Write the information exactly the way it is listed on the care label and indicate if the garment is in compliance for each of the areas that follow.

Labeling Content Evaluation

	Label Content	Compliant
Washing Instructions: Temperature and cycle, any modifiers (i.e., gentle cycle, hand wash, turn inside out, wash separately)		☐Yes ☐No
Bleaching instructions		☐Yes ☐No
Drying instructions: Temperature and cycle or other method (i.e., line dry, dry flat)		☐Yes ☐No
Ironing instructions: Temperature and any other modifiers (i.e., no steam, turn inside out)		☐Yes ☐No
Dryclean instructions		☐Yes ☐No
Warning instructions		☐Yes ☐No

Lab Activity 7.2 *(cont.)*

Are the care instructions complete and compliant? ☐Yes ☐No If no, how would your team correct them.

Are care symbols used on the label? ☐Yes ☐No If no, draw or insert the garment symbols that correspond with the written care instructions on the label. Refer to Figures 7.1–7.3 in the text for ASTM, ISO/GINETEX, or JIS care symbols.

Lab Activity 7.2 *(cont.)*

Does your team believe this is the best care method for this garment? ☐Yes ☐No If no, rewrite the care instructions to reflect your groups recommended care method.

NOTES

CHAPTER 8

Safety Regulations and Guidelines for Wearing Apparel Lab

LAB OBJECTIVES:

- To examine garment for compliance with safety regulations.

Chapter 10 of *Apparel Quality: A Guide to Evaluating Sewn Products* focuses on the mandatory safety regulations and voluntary guidelines for apparel sold in the United States, Canada, the European Union, and Japan, as well as the regulatory bodies overseeing and enforcing safety regulations. The goal of brands around the globe is to design and manufacture apparel products that are compliant with relevant governmental regulations to avoid causing intentional harm to consumers. Education of industry professionals is vital to ensuring products are developed and tested to certify they are safe for the useful life of the product.

There is one lab activity (worksheet 8.1) for this chapter, and two additional activities that are carried over from Chapters 3 and 4 (Lab Activities 3.5 and 4.3 Comparison Project garment analysis after week 5 of the wear test). The lab activity for this chapter will focus on evaluating garments for safety to determine compliance with safety regulations in the United States, Canada, the European Union, and Japan.

PREPARATION AND SUPPLIES FOR CHAPTER 8 LAB ACTIVITIES

Prior to Chapter 8 lab activities, wear your comparison project garment for the designated number of hours and clean it according to the garment label care instructions. Both the wear test garment and the control must be brought to lab for evaluation.

Lab Activity 8.1: Garment Safety Regulations and Compliance

For Lab Activity 8.1: Garment Safety Regulations and Compliance, work in teams of two to three people. Each group will collaboratively evaluate the children's apparel items in worksheet 8.1. Analyze each garment for both design and materials selection in relation to the safety of the intended user to determine compliance with safety regulations for the products to enter into commerce in the United States, Canada, the European Union, and Japan. Refer to Table 8.1 for a summary of safety regulations by country.

Table 8.1 Summary of Safety Regulations by Country

REGULATION	COUNTRY			
	U.S.	CANADA	HARMONIZED EU	JAPAN
Nickel	N	N	R	N
Lead	R	N	N	N
Certain Azo Dyes	N	N	R	N
Formaldehyde	R	N	N	R
Fluorinated Organic Compounds	V	R	R	N
Certain Flame Retardants	R	R	R	R
Flammability - General Wearing Apparel	R	R	N	N
Flammability - Children's Sleepwear	R	R	R	N
Drawstrings in Children's Apparel	V	V	R	N

Note:

R = Required Regulation

V = Voluntary Regulation

N = No Regulation to Date

Lab Activity 3.5 Continued: Wear-Testing Comparison Project Garment

Go back to the Lab Activity 3.5: Wear-Testing Comparison Project Garment worksheet. Update your wear log and refurbishment method for week 5. Measure the garment according to the specified method used, when you originally measured the garment. Calculate dimensional change for week 5, and record the results on worksheet 3.5. After you have completed this task, proceed to Lab Activity 4.3: Comparison Project Garment Evaluation of Appearance and Color Change after Laundering.

Lab Activity 4.3 Continued: Comparison Project Garment Evaluation of Appearance and Color Change after Laundering

For Lab Activity 4.3, each group member will bring his or her Comparison Project garments to lab for evaluation. Each team member will compare his or her control garment to the wear-tested garment for week 5 to determine if changes in color and appearance have occurred or if the garment has remained the same. Document your individual findings for week 5 on worksheet 4.3.

Lab Activity 8.1: GARMENT SAFETY REGULATIONS AND COMPLIANCE

Evaluate each of the garments listed here and answer the questions regarding safety compliance. Use Table 8.1 to help you complete this worksheet.

Your Name

Name(s) of Teammate(s)

Children's Sweatshirt with Hood	
Available in sizes 2T to 6x 100% cotton Made in Cambodia Pullover fleece sweatshirt with drawstring hood and pouch pocket	

This garment will be shipped to retail stores in the United States, European Union, Canada, and Japan. Are there any safety regulations to be concerned with for the countries listed? ☐Yes ☐No If yes, list the safety concern(s)?

Lab Activity 8.1 *(cont.)*

Indicate if the garment is compliant with safety regulations for each of these countries/territories. Note if the safety concern is regulated and if it is, indicate if it is voluntary or required?					
United States	☐Yes	☐No	☐Voluntary	☐Required	☐Not Regulated
European Union	☐Yes	☐No	☐Voluntary	☐Required	☐Not Regulated
Canada	☐Yes	☐No	☐Voluntary	☐Required	☐Not Regulated
Japan	☐Yes	☐No	☐Voluntary	☐Required	☐Not Regulated

If you answered no for any of these countries, explain how you would redesign the garment to make it compliant with safety regulations to be sold in all of these territories.

Lab Activity 8.1 *(cont.)*

Children's Jeans	
Available in sizes 5 to 14 100% cotton Made in China Denim fly front zipper jean with nickel tack button and elastic back waistband	
This garment will be shipped to retail stores in the United States, European Union, Canada, and Japan. Are there any safety regulations to be concerned with for the countries listed? ☐Yes ☐No If yes, list the safety concern(s)?	

Lab Activity 8.1 *(cont.)*

Indicate if the garment is compliant with safety regulations for each of these countries/territories. Note if the safety concern is regulated and if it is, indicate if it is voluntary or required?					
United States	☐Yes	☐No	☐Voluntary	☐Required	☐Not Regulated
European Union	☐Yes	☐No	☐Voluntary	☐Required	☐Not Regulated
Canada	☐Yes	☐No	☐Voluntary	☐Required	☐Not Regulated
Japan	☐Yes	☐No	☐Voluntary	☐Required	☐Not Regulated

If you answered no for any of these countries, explain how you would redesign the garment to make it compliant with safety regulations to be sold in all of these territories.

Lab Activity 8.1 *(cont.)*

Children's Pajamas	
Available in sizes 4 to 6X 100% cotton treated with TRIS flame retardant Jersey knit pajamas with ribbon tied into a bow at the neckline and elastic waist pull-on pants	
This garment will be shipped to retail stores in the United States, European Union, Canada, and Japan. Are there any safety regulations to be concerned with for the countries listed? ☐Yes ☐No If yes, list the safety concern(s)?	

Lab Activity 8.1 *(cont.)*

Indicate if the garment is compliant with safety regulations for each of these countries/territories. Note if the safety concern is regulated and if it is, indicate if it is voluntary or required?					
United States	☐Yes	☐No	☐Voluntary	☐Required	☐Not Regulated
European Union	☐Yes	☐No	☐Voluntary	☐Required	☐Not Regulated
Canada	☐Yes	☐No	☐Voluntary	☐Required	☐Not Regulated
Japan	☐Yes	☐No	☐Voluntary	☐Required	☐Not Regulated

If you answered no for any of these countries, explain how you would redesign the garment to make it compliant with safety regulations to be sold in all of these territories.

Illustrations by Janace Bubonia

NOTES

CHAPTER 9

Measuring Quality through Raw Materials and Product Testing Lab

LAB OBJECTIVES:

- To conduct tests to evaluate product performance and appearance.
- To utilize a standard product performance specification to determine if a product is meeting minimum quality standards based on test data collected.

Chapters 11 and 12 of *Apparel Quality: A Guide to Evaluating Sewn Products* focus on the importance of standard test methods to provide a means for companies to obtain reliable, reproducible results regarding the materials and garments they are producing and selling. A garment's appearance and performance plays a major role in customer satisfaction and perception of quality. Testing and evaluating materials and finished garments can allow for improvements to be made to increase customer satisfaction and avoid unnecessary returns due to poor performance. When apparel products are tested and analyzed for performance, function, safety, and appearance, this leads to better quality garments and ultimately higher customer satisfaction.

There are four lab activities (worksheets 9.1, 9.2, 9.3, 9.4) for this chapter, which focus on performance and appearance testing for piece goods, the Comparison Project physical testing garment, and selected garments from your group's wardrobe. Lab Activities 9.2 and 9.3 may take several lab sessions to complete all of the testing. After you have completed Lab Activity 9.3, you can proceed with Lab Activity 9.4.

PREPARATION AND SUPPLIES FOR CHAPTER 9 LAB ACTIVITIES

For Chapter 9 lab activities, bring the piece good specimens and the comparison project garment for physical testing as well as, a ruler, scissors, a permanent marking pen, and at least 8 pages of card stock for mounting specimens after testing. Lab activities for Chapter 9 will span several sessions so be sure to bring all of the supplies and test specimens to lab.

Lab Activity 9.1: Test Methods for Evaluating Selected Garments

For Lab Activity 9.1: Test Methods for Evaluating Selected Garments, work in teams of two to three people. Each team member will bring a garment of his or her choice. As a team, determine which tests should be conducted to evaluate fabric characterization, garment characteristics for appearance, performance, and safety aspects. Select the standard test methods to be used from ASTM International (ASTM), American Association of Textile Chemists and Colorists (AATCC), International Organization for Standardization (ISO), British Standards Institution (BSI), European Committee for Standardization (CEN), or Japanese Industrial Standards (JIS) that correspond with the types of testing to be conducted from Tables 9.1, 9.2, 9.3, and 9.4. Explain what each test method is used to measure.

Lab Activity 9.2: Piece Goods Appearance and Performance Testing

For Lab Activity 9.2: Piece Goods Appearance and Performance Testing, work in teams of four to five people to conduct appearance and performance testing on your 2-yard/-meter lab sample of piece goods. Before you begin testing, review the test standards to ensure you have fully and accurately prepared the specimens for testing. Mount tested specimens on cardstock and attach then to worksheet 9.2.

Lab Activity 9.3: Comparison Garment Appearance and Performance Testing

For Lab Activity 9.3: Comparison Garment Appearance and Performance Testing, work in teams of four to five people to prepare specimens and conduct appearance and performance testing on your Comparison Project garment designated for physical testing. Mark and cut specimens from the comparison shirt. Use the following templates prepared for Lab Activity 3.2: crocking, pilling resistance, and bursting strength. Your instructor may have required additional templates to be prepared for comparison testing, depending upon the testing equipment, apparatuses, and available test standards. Mark the Comparison Project garment designated for physical testing where the specimens will be taken. It is important to place the templates close together to utilize the sample wisely. Lay out and cut specimens for the same test by staggering them in different areas of the Comparison Project sample. Mark all specimens in the warp direction with a small line away from the testing area. Before testing begins, review the test standards to ensure you have fully and accurately prepared the specimens. Mount tested specimens to cardstock and attach them to worksheet 9.3.

Lab Activity 9.4: Comparison Project Garment Results and Performance Specifications

For Lab Activity 9.4: Comparison Project Garment Results and Performance Specifications, you will independently enter your test results into worksheet 9.4 along with the performance specifications from the designated standard. Determine if the Comparison Project garment has passed or failed each of the tests based on the minimum or maximum performance specifications.

Table 9.1 Test Methods Commonly Used for Evaluating Dimensional Change of Textile Materials or Garments

TEST METHOD DESIGNATION	NAME
AATCC 96	Dimensional Changes in Commercial Laundering of Woven and Knitted Fabrics except Wool
AATCC 135	Dimensional Changes of Fabrics after Home Laundering
AATCC 150	Dimensional Changes of Garments after Home Laundering
AATCC 158	Dimensional Changes on Drycleaning in Perchloroethylene: Machine Method
AATCC 187	Dimensional Changes of Fabrics: Accelerated
ASTM 6207	Dimensional Stability of Fabrics to Changes in Humidity and Temperature
EN ISO 5077	Textiles—Determination of Dimensional Change in Washing and Drying
EN ISO 3175-1	Textiles—Professional Care, Drycleaning and Wetcleaning of Fabrics and Garments—Part 1: Assessment of Performance after Cleaning and Finishing
EN ISO 3175-2	Textiles—Professional Care, Drycleaning and Wetcleaning of Fabrics and Garments—Part 2: Procedure for Testing Performance when Cleaning and Finishing using Tetrachloroethylene
EN ISO 3175-3	Textiles—Professional Care, Drycleaning and Wetcleaning of Fabrics and Garments—Part 3: Procedure for Testing Performance when Cleaning and Finishing using Hydrocarbon Solvents
EN ISO 3175-4	Textiles—Professional Care, Drycleaning and Wetcleaning of Fabrics and Garments—Part 4: Procedure for Testing Performance when Cleaning and Finishing using Simulated Wetcleaning
EN ISO 3759	Textiles—Preparation, Marking and Measuring of Fabric Specimens and Garments in Tests for Determination of Dimensional Change
EN ISO 6330	Textiles—Domestic Washing and Drying Procedures for Textile Testing
EN ISO 23231	Textiles—Determination of Dimensional Change of Fabrics—Accelerated Machine Method
JIS L 1909	Textiles—Determination of Dimensional Change

Sources: ASTM International, "D6322-07 Standard Guide to International Test Methods Associated with Textile Care Procedures," *2014 Annual Book of ASTM Standards* (West Conshohocken, PA: ASTM International, 2014); AATCC, *2013 Technical Manual of the American Association of Textile Chemists and Colorists*, vol. 88 (Research Triangle Park, NC: AATCC, 2013); ISO, *BS EN ISO 5077:2008, Textiles—Determination of Dimensional Change in Washing and Drying*, 1–2, Geneva, CE: ISO, 2014; JSA Web Store, http://www.webstore.jsa.or.jp/webstore/Top/indexEn.jsp, accessed April 7, 2013; CEN, European Committee for Standardization, Catalogue of Published Standards, http://esearch.cen.eu/esearch/extendedsearch.aspx, accessed April 7, 2013.

Table 9.2 Test Methods Commonly Used for Evaluating Appearance of Textile Materials or Garments

TEST METHOD DESIGNATION	NAME
AATCC 66	Wrinkle Recovery of Woven Fabrics Recovery Angle
AATCC 88B	Smoothness of Seams in Fabrics after Repeated Home Laundering
AATCC 88C	Retention of Creases in Fabrics after Repeated Home Laundering
AATCC 124	Smoothness Appearance of Fabrics after Repeated Home Laundering
AATCC 128	Wrinkle Recovery of Fabrics: Appearance Method
AATCC 143	Appearance of Apparel and Other Textile End Products after Repeated Home Laundering
EN ISO 2313	Textiles—Determination of the Recovery from Creasing of a Horizontally Folded Specimen of Fabric by Measuring the Angle of Recovery
EN ISO 7768	Textiles—Test Method for Assessing Smoothness Appearance of Fabrics after Cleansing
EN ISO 7769	Textiles—Test Method for Assessing the Appearance of Creases in Fabrics after Cleansing
EN ISO 7770	Textiles—Test Method for Assessing the Smoothness Appearance of Seams in Fabrics after Cleansing
EN ISO 9867	Evaluation of the Wrinkle Recovery of Fabrics—Appearance Methods
JIS L 1059-1	Testing Methods for Crease Recovery of Textiles—Part 1: Determination of the Recovery from Creasing of a Horizontally Folded Specimen by Measuring the Angle of Recovery
JIS L 1059-2	Testing Methods for Crease Recovery of Textiles—Part 2: Evaluation of the Wrinkle Recovery of Fabrics—Appearance Method
JIS L 1905	Methods for Assessing the Appearance of Seam Pucker on Textiles

Sources: ASTM International, "D6322-07 Standard Guide to International Test Methods Associated with Textile Care Procedures", *2014 Annual Book of ASTM Standards* (West Conshohocken, PA: ASTM International, 2014); AATCC, *2013 Technical Manual of the American Association of Textile Chemists and Colorists*, vol. 88 (Research Triangle Park, NC: AATCC, 2013); ISO, *BS EN ISO 5077:2008, Textiles—Determination of Dimensional Change in Washing and Drying*, Brussels, BE: CEN European Committee for Standardization, 20141–2; ANSI eStandards store, http://webstore.ansi.org/default.aspx, accessed July 29, 2013; JSA Web Store, http://www.webstore.jsa.or.jp/webstore/Top/indexEn.jsp, accessed April 7, 2013; CEN, European Committee for Standardization, Catalogue of Published Standards, http://esearch.cen.eu/esearch/extendedsearch.aspx, accessed April 7, 2013.

Table 9.3 Test Methods Commonly Used for Evaluating Color Permanence of Textile Materials or Garments

TEST METHOD DESIGNATION	NAME
AATCC 8	Colorfastness to Crocking
AATCC 15	Colorfastness to Perspiration
AATCC 16.1	Colorfastness to Light: Outdoor
AATCC 16.2	Colorfastness to Light: Carbon-Arc
AATCC 16.3	Colorfastness to Light: Xenon-Arc
AATCC 23	Colorfastness to Burn Gas Fumes
AATCC 61	Colorfastness to Laundering: Accelerated
AATCC 101	Colorfastness to Bleaching with Hydrogen Peroxide
AATCC 104	Colorfastness to Water Spotting
AATCC 106	Colorfastness to Water: Sea
AATCC 107	Colorfastness to Water
AATCC 109	Colorfastness to Ozone in the Atmosphere under Low Humidities
AATCC 110	Whiteness of Textiles
AATCC 116	Colorfastness to Crocking: Rotary Vertical Crockmeter Method
AATCC 117	Colorfastness to Heat: Dry (Excluding Pressing)
AATCC 125	Colorfastness to Perspiration and Light
AATCC 129	Colorfastness to Ozone in the Atmosphere under High Humidities
AATCC 131	Colorfastness to Pleating: Steam Pleating
AATCC 132	Colorfastness to Drycleaning
AATCC 133	Colorfastness to Heat: Hot Pressing
AATCC 157	Colorfastness to Solvent Spotting: Perchloroethylene
AATCC 162	Colorfastness to Water: Chlorinated Pool
AATCC 163	Colorfastness: Dye Transfer in Storage: Fabric-to-Fabric

continued

Table 9.3 *(cont.)*

TEST METHOD DESIGNATION	NAME
AATCC 164	Colorfastness to Oxides of Nitrogen in Atmosphere under High Humidities
AATCC 172	Colorfastness to Powdered Non-Chlorine Bleach in Home Laundering
AATCC 188	Colorfastness to Sodium Hypochlorite Bleach in Home Laundering
AATCC 190	Colorfastness to Home Laundering with Activated Oxygen Bleach Detergent: Accelerated
ASTM D2052	Colorfastness of Zippers to Drycleaning
ASTM D2053	Colorfastness of Zippers to Light
ASTM D2054	Colorfastness of Zipper Tapes to Crocking
ASTM D2057	Colorfastness of Zippers to Laundering
EN ISO 105-A01	Textiles—Test for Colour Fastness—Part A01: General Principles of Testing
EN ISO 105-A02	Textiles—Test for Colour Fastness—Part A02: Grey Scale for Assessing Change in Colour
EN ISO 105-A03	Textiles—Test for Colour Fastness—Part A03: Grey Scale for Assessing Staining
EN ISO 105-A04	Textiles—Test for Colour Fastness—Part A04: Method for the Instrumental Assessment of the Degree of Staining of Adjacent Fabrics
EN ISO 105-A05	Textiles—Test for Colour Fastness—Part A05: Instrumental Assessment of Change in Colour for Determination of Grey Scale Rating
EN ISO 105-A11	Textiles—Test for Colour Fastness—Part A11: Determination of Colour Fastness Grades by Digital Imaging Techniques
EN ISO 105-B01	Textiles—Test for Colour Fastness—Part B01: Colour Fastness to Light: Daylight
EN ISO 105-B02	Textiles—Test for Colour Fastness—Part B02: Colour Fastness to Artificial Light: Xenon Arc Fading Lamp Test
EN ISO 105-B03	Textiles—Test for Colour Fastness—Part B03: Colour Fastness to Weathering: Outdoor Exposure
EN ISO 105-B04	Textiles—Test for Colour Fastness—Part B04: Colour Fastness to Artificial Weathering: Xenon Arc Fading Lamp Test
EN ISO 105-B06	Textiles—Test for Colour Fastness—Part B06: Colour Fastness and Aging to Artificial Light at High Temperatures: Xenon Arc Fading Lamp Test
EN ISO 105-B07	Textiles—Test for Colour Fastness—Part B07: Colour Fastness to Light of Textiles Wetted with Artificial Perspiration
EN ISO 105-B10	Textiles—Test for Colour Fastness—Part B10: Artificial Weathering—Exposure to Filtered Xenon-Arc Radiation

Table 9.3 *(cont.)*

TEST METHOD DESIGNATION	NAME
EN ISO 105-C06	Textiles—Test for Colour Fastness—Part C06: Colour Fastness to Domestic and Commercial Laundering
EN ISO 105-C08	Textiles—Test for Colour Fastness—Part C08: Colour Fastness to Domestic and Commercial Laundering Using a Non-Phosphate Reference Detergent Incorporating a Low-Temperature Bleach Activator
EN ISO 105-C09	Textiles—Test for Colour Fastness—Part C09: Colour Fastness to Domestic and Commercial Laundering—Oxidative Bleach Response Using a Non-Phosphate Reference Detergent Incorporating a Low-Temperature Bleach Activator
EN ISO 105-C10	Textiles—Test for Colour Fastness—Part C10: Colour Fastness to Washing with Soap or Soap and Soda
EN ISO 105-D01	Textiles—Test for Colour Fastness—Part D 01: Colour Fastness to Drycleaning Using Perchloroethylene Solvent
EN ISO 105-D02	Textiles—Test for Colour Fastness—Part D 02: Colour Fastness to Rubbing: Organic Solvents
EN ISO 105-E01	Textiles—Test for Colour Fastness—Part E 01: Colour Fastness to Water
EN ISO 105-E02	Textiles—Test for Colour Fastness—Part E 02: Colour Fastness to Sea Water
EN ISO 105-E03	Textiles—Test for Colour Fastness—Part E 03: Colour Fastness to Chlorinated Water (Swimming-pool Water)
EN ISO 105-E04	Textiles—Test for Colour Fastness—Part E 04: Colour Fastness to Perspiration
EN ISO 105-E05	Textiles—Test for Colour Fastness—Part E 05: Colour Fastness to Spotting: Acid
EN ISO 105-E06	Textiles—Test for Colour Fastness—Part E 06: Colour Fastness to Spotting: Alkali
EN ISO 105-E07	Textiles—Test for Colour Fastness—Part E 07: Colour Fastness to Spotting: Water
EN ISO 105-E08	Textiles—Test for Colour Fastness—Part E 08: Colour Fastness to Hot Water
EN ISO 105-E11	Textiles—Test for Colour Fastness—Part E 11: Colour Fastness to Steaming
EN ISO 105-G01	Textiles—Test for Colour Fastness—Part G 01: Colour Fastness to Nitrogen Oxides
EN ISO 105-G02	Textiles—Test for Colour Fastness—Part G 02: Colour Fastness to Burnt-Gas Fumes
EN ISO 105-G03	Textiles—Test for Colour Fastness—Part G 03: Colour Fastness to Ozone in the Atmosphere
EN ISO 105-G04	Textiles—Test for Colour Fastness—Part G 04: Colour Fastness to Oxides of Nitrogen in the Atmosphere at High Temperatures
EN ISO 105-J01	Textiles—Test for Colour Fastness—Part J 01: Colour Fastness to Ozone in the Atmosphere

continued

Table 9.3 *(cont.)*

TEST METHOD DESIGNATION	NAME
EN ISO 105-J02	Textiles—Test for Colour Fastness—Part J 02: Instrumental Assessment of Relative Whiteness
EN ISO 105-J03	Textiles—Test for Colour Fastness—Part J 03: Calculation of Color Differences
EN ISO 105-N01	Textiles—Test for Colour Fastness—Part N 01: Color Fastness to Bleaching: Hypochlorite
EN ISO 105-N02	Textiles—Test for Colour Fastness—Part N 02: Color Fastness to Bleaching: Peroxide
EN ISO 105-N03	Textiles—Test for Colour Fastness—Part N 03: Color Fastness to Bleaching: Sodium Chlorite (mild)
EN ISO 105-N04	Textiles—Test for Colour Fastness—Part N 04: Color Fastness to Bleaching: Sodium Chlorite (severe)
EN ISO 105-P01	Textiles—Test for Colour Fastness—Part P 01: Color Fastness to Pleating: Steam Pleating
EN ISO 105-X05	Textiles—Test for Colour Fastness—Part X 05: Color Fastness to Organic Solvents
EN ISO 105-X11	Textiles—Test for Colour Fastness—Part X 11: Color Fastness to Hot Pressing
EN ISO 105-X12	Textiles—Test for Colour Fastness—Part X 12: Color Fastness to Rubbing
JIS L 0801	General Principles of Testing Methods for Colour Fastness
JIS L 0803	Standard Adjacent Fabrics for Staining of Colour Fastness Test
JIS L 0841	Test Methods for Colour Fastness to Daylight
JIS L 0842	Test Methods for Colour Fastness to Enclosed Carbon Arc Lamp Light
JIS L 0843	Test Methods for Colour Fastness to Xenon Arc Lamp Light
JIS L 0844	Test Methods for Colour Fastness to Washing and Laundering
JIS L 0845	Test Methods for Colour Fastness to Hot Water
JIS L 0846	Test Methods for Colour Fastness to Water
JIS L 0847	Test Methods for Colour Fastness to Sea Water
JIS L 0848	Test Methods for Colour Fastness to Perspiration
JIS L 0849	Test Methods for Colour Fastness to Rubbing
JIS L 0850	Test Methods for Colour Fastness to Hot Pressing
JIS L 0851	Test Methods for Colour Fastness to Acid Spotting
JIS L 0852	Test Methods for Colour Fastness to Alkali Spotting

Table 9.3 *(cont.)*

TEST METHOD DESIGNATION	NAME
JIS L 0853	Test Methods for Colour Fastness to Water Spotting
JIS L 0854	Test Methods for Colour Fastness to Sublimation in Storage
JIS L 0856	Test Methods for Colour Fastness to Bleaching with Hypochlorite
JIS L 0857	Test Methods for Colour Fastness to Bleaching with Peroxide
JIS L 0859	Test Methods for Colour Fastness to Bleaching with Sodium Chlorite
JIS L 0860	Test Methods for Colour Fastness to Dry Cleaning
JIS L 0861	Test Methods for Colour Fastness to Organic Solvents
JIS L 0862	Test Methods for Colour Fastness to Rubbing with Organic Solvents
JIS L 0869	Test Methods for Colour Fastness to Steaming
JIS L 0873	Test Methods for Colour Fastness to Chlorination
JIS L 0878	Test Methods for Colour Fastness to Stem Pleating
JIS L 0879	Test Methods for Colour Fastness to Dry Heat
JIS L 0880	Test Methods for Colour Fastness to Pleating
JIS L 0884	Test Methods for Colour Fastness to Chlorinated Water
JIS L 0888	Test Methods for Colour Fastness to Light and Perspiration
JIS L 0889	Test Methods for Colour Fastness to Bleaching Laundering with Sodium Percarbonate
JIS L 0890	Test Methods for Colour Fastness to Ozone
JIS L 0891	Test Methods for Colour Fastness to Artificial Accelerated Weathering with Xenon Arc Lamp Light or Sunshine Carbon Arc Lamp Light
JIS L 1916	Determination of Whiteness for Textiles

Sources: ASTM International, "D6322-07 Standard Guide to International Test Methods Associated with Textile Care Procedures," *2014 Annual Book of ASTM Standards* (West Conshohocken, PA: ASTM International, 2014); AATCC, *2013 Technical Manual of the American Association of Textile Chemists and Colorists*, vol. 88 (Research Triangle Park, NC: AATCC, 2013); *2014 Annual Book of ASTM Standards* (West Conshohocken, PA: ASTM International, 2014); ISO, *BS EN ISO 5077:2008, Textiles—Determination of Dimensional Change in Washing and Drying*,Brussels: BE: CEN European Committee for Standardization, 2014 1–2; ANSI eStandards store, http://webstore.ansi.org/default.aspx, accessed July 29, 2013; JSA Web Store, http://www.webstore.jsa.or.jp/webstore/JIS/FlowControl.jsp, accessed April 7, 2013; CEN, European Committee for Standardization, Catalogue of Published Standards, http://esearch.cen.eu/esearch/extendedsearch.aspx, accessed April 7, 2013.

Table 9.4 Test Methods Commonly Used for Evaluating Abrasion and Durability of Textile Materials or Garments

TEST METHOD DESIGNATION	NAME
AATCC 93	Abrasion Resistance of Fabrics Accelerator Method
AATCC 136	Bond Strength of Bonded and Laminated Fabrics
ASTM D1336	Standard Test Method for Distortion of Yarn in Woven Fabrics
ASTM D1424	Standard Test Method for Tearing Strength of Fabrics by Falling-Pendulum (Elmendorf-Type) Apparatus
ASTM D1683	Standard Test Method for Failure of Sewn Seams of Woven Apparel Products
ASTM D2261	Standard Test Method for Tearing Strength of Fabrics by the Tongue (Single Rip) Procedure (Constant-Rate-of -Extension Tensile Testing Machine)
ASTM D2594	Standard Test Method for Stretch Properties of Knitted Fabrics Having Low Power
ASTM D2724	Standard Test Method for Bonded, Fused, and Laminated Apparel Fabrics
ASTM D3107	Standard Test Method for Stretch Properties of Fabrics Woven from Stretch Yarns
ASTM D3511	Standard Test Method for Pilling Resistance and Other Related Surface Changes of Textile Fabrics: Brush Pilling Tester
ASTM D3512	Standard Test Method for Pilling Resistance and Other Related Surface Changes of Textile Fabrics: Random Tumble Pilling Tester
ASTM D3514	Standard Test Method for Pilling Resistance and Other Related Surface Changes of Textile Fabrics: Elastomeric Pad
ASTM D3786	Standard Test Method for Bursting Strength of Textile
	Fabrics—Diaphragm Bursting Strength Tester Method
ASTM D3787	Standard Test Method for Bursting Strength of Textile
	Fabrics—Constant-Rate-of-Transverse (CRT) Ball Burst Test
ASTM D3884	Standard Test Method for Abrasion Resistance of Textile Fabrics (Rotary Platform, Double-Head Method)
ASTM D3885	Standard Test Method for Abrasion Resistance of Textile Fabrics (Flexing and Abrasion Method)
ASTM D3886	Standard Test Method for Abrasion Resistance of Textile Fabrics (Inflated Diaphragm Apparatus)
ASTM D3939	Standard Test Method for Snagging Resistance of Fabrics (Mace)

Table 9.4 *(cont.)*

TEST METHOD DESIGNATION	NAME
ASTM D4157	Standard Test Method for Abrasion Resistance of Textile Fabrics (Oscillatory Cylinder Method)
ASTM D4685	Standard Test Method for Pile Fabric Abrasion
ASTM D4964	Standard Test Method for Tension and Elongation of Elastic Fabrics (Constant-Rate-of-Extension Type Tensile Testing Machine)
ASTM D4966	Standard Test Method for Abrasion Resistance of Textile Fabrics (Martindale Abrasion Tester Method)
ASTM D4970	Standard Test Method for Pilling Resistance and Other Related Surface Changes of Textile Fabrics: Martindale Tester
ASTM D5034	Standard Test Method for Breaking Strength and Elongation of Textile Fabrics (Grab Test)
ASTM D5035	Standard Test Method for Breaking Force and Elongation of Textile Fabrics (Strip Method)
ASTM D5278	Elongation of Narrow Elastic Fabrics (Static Load Testing)
ASTM D5362	Standard Test Method for Snagging Resistance of Fabrics (Bean Bag)
ASTM D5587	Standard Test Method for Tearing Strength by Trapezoid Procedure
ASTM D6614	Standard Test Method for Stretch Properties of Textile Fabrics—CRE Method
ASTM D6770	Standard Test Method for Abrasion Resistance of Textile Fabrics (Hex Bar Method)
ASTM D6775	Standard Test Method for Breaking Strength and Elongation of Textile Webbing, Tape, and Braided Material
ASTM D6797	Standard Test Method for bursting Strength of Fabrics Constant-Rate-of-Extension (CRE) Ball Burst
EN ISO 9073-4	Textiles—Test Methods for Nonwovens—Part 4: Determination of Tear Resistance
EN ISO 9073-5	Textiles—Test Methods for Nonwovens—Part 5: Determination of Resistance to Mechanical Penetration (Ball Burst Procedure)
EN ISO 9073-18	Textiles—Test Methods for Nonwovens—Part 18: Determination of Breaking Strength and Elongation of Nonwoven Materials Using the Grab Tensile Test
EN ISO 12945-1	Textiles—Determination of Fabric Propensity to Surface Fuzzing and to Pilling—Part 1: Pilling Box Method
EN ISO 12945-2	Textiles—Determination of Fabric Propensity to Surface Fuzzing and to Pilling—Part 2: Modified Martindale Method

continued

Table 9.4 *(cont.)*

TEST METHOD DESIGNATION	NAME
EN ISO 12947-1	Textiles—Determination of the Abrasion Resistance of Fabrics by the Martindale Method—Part 1: Martindale Abrasion Testing Apparatus
EN ISO 12947-2	Textiles—Determination of the Abrasion Resistance of Fabrics by the Martindale Method—Part 2: Determination of Specimen Breakdown
EN ISO 12947-3	Textiles—Determination of the Abrasion Resistance of Fabrics by the Martindale Method—Part 3: Determination of Mass Loss
EN ISO 12947-4	Textiles—Determination of the Abrasion Resistance of Fabrics by the Martindale Method—Part 4: Assessment of Appearance Change
EN ISO 13770	Textiles—Determination of the Abrasion Resistance of Knitted Footwear Garments
EN ISO 13934-1	Textiles—Tensile Properties of Fabrics—Part 1: Determination of Maximum Force and Elongation at Maximum Force Using the Strip Method
EN ISO 13934-2	Textiles—Tensile Properties of Fabrics—Part 2: Determination of Maximum Force and Elongation Using the Grab Method
EN ISO 13935-1	Textiles—Seam Tensile Properties of Fabrics and Made-up Textile Articles—Part 1: Determination of Maximum Force to Seam Rupture Using the Strip Method
EN ISO 13935-2	Textiles—Seam Tensile Properties of Fabrics and Made-up Textile Articles—Part 2: Determination of Maximum Force to Seam Rupture Using the Grab Method
EN ISO 13936-1	Textiles—Determination of the Slippage Resistance of Yarns at a Seam in Woven Fabrics—Part 1: Fixed Seam Opening Method
EN ISO 13936-2	Textiles—Determination of the Slippage Resistance of Yarns at a Seam in Woven Fabrics—Part 3: Needle Clamp Method
EN ISO 13936-3	Textiles—Determination of the Slippage Resistance of Yarns at a Seam in Woven Fabrics—Part 1: Fixed Seam Opening Method
EN ISO 13937-1	Textiles—Tear Properties of Fabrics—Part 1: Determination of Tear Force Using Ballistic Pendulum Method
EN ISO 13937-2	Textiles—Tear Properties of Fabrics—Part 2: Determination of Tear Force of Trouser-Shaped Test Specimens (Single Tear Method)
EN ISO 13937-3	Textiles—Tear Properties of Fabrics—Part 3: Determination of Tear Force of Wing-Shaped Test Specimens (Single Tear Method)
EN ISO 13937-4	Textiles—Tear Properties of Fabrics—Part 4: Determination of Tear Force of Tongue-Shaped Test Specimens (Double Tear Test)
EN ISO 13938-1	Textiles—Bursting Strength Properties of Fabrics—Part 1: Hydraulic Method of Determination of Bursting Strength and Bursting Distension

Table 9.4 *(cont.)*

TEST METHOD DESIGNATION	NAME
EN ISO 14704-1	Determination of Elasticity of Fabrics—Part 1: Strip Method
EN ISO 14704-2	Determination of Elasticity of Fabrics—Part 2: Multiaxial Tests
EN ISO 14704-3	Determination of Elasticity of Fabrics—Part 3: Narrow Fabrics
EN ISO 29073	Textiles—Test Methods for Nonwovens—Part 3: Determination of Tensile Strength and Elongation
JIS L 1061	Test Methods for Bagging of Woven and Knitted Fabrics
JIS L 1062	Test Methods for Distortion and Slippage of Yarn in Woven Fabrics
JIS L 1075	Testing Methods for Pile Retention of Woven and Knitted Fabrics
JIS L 1076	Test Methods for Pilling of Woven and Knitted Fabrics
JIS L 1093	Test Methods for Seam Strength of Textiles
JIS L 1901	Test Methods for Frosting Due to Yarn Reversing of Woven or Knitted Fabrics
JIS L 1905	Test Methods for Assessing Appearance of Seam Pucker on Textiles
JIS L 1910	Test Methods for Percentage of Breaking Strength and Bursting Strength Lowering of Textiles to Oxygen Bleaching

Sources: ASTM International, *2014 Annual Book of ASTM Standards* (West Conshohocken, PA: ASTM International, 2014); AATCC, *2013 Technical Manual of the American Association of Textile Chemists and Colorists*, vol. 88 (Research Triangle Park, NC: AATCC, 2013); ANSI eStandards store, http://webstore.ansi.org/default.aspx, accessed July 29, 2013; JSA Web Store, http://www.webstore.jsa.or.jp/webstore/JIS/FlowControl.jsp, accessed April 7, 2013; CEN, European Committee for Standardization, Catalogue of Published Standards, http://esearch.cen.eu/esearch/extendedsearch.aspx, accessed April 7, 2013.

Lab Activity 9.1: TEST METHODS FOR EVALUATING SELECTED GARMENTS

Evaluate each of the garments as a group and record the information for your garment in this worksheet. Determine which test methods would be best suited for evaluating the products for fabric characterization, appearance, performance, and safety aspects. Use Tables 9.1, 9.2, 9.3, and 9.4 to complete this worksheet.

Your Name

Name(s) of Teammate(s)

Product Description

Test Method Number and Name for Evaluation of Fiber Identification
Explanation of What the Test Measures

Lab Activity 9.1 *(cont.)*

Test Method Number and Name for Evaluation of Yarn Construction
Explanation of What the Test Measures

Test Method Number and Name for Evaluation of Fabric Count
Explanation of What the Test Measures

Lab Activity 9.1 *(cont.)*

Test Method Number and Name for Evaluation of Fabric Weight
Explanation of What the Test Measures

Test Method Number and Name for Evaluation of Dimensional Change
Explanation of What the Test Measures

Lab Activity 9.1 *(cont.)*

Test Method Number and Name for Evaluation of Appearance
Explanation of What the Test Measures

Test Method Number and Name for Evaluation of Color Permanence
Explanation of What the Test Measures

Lab Activity 9.1 *(cont.)*

Test Method Number and Name for Evaluation of Color Permanence (different from that listed previously)
Explanation of What the Test Measures

Test Method Number and Name for Evaluation of Color Permanence (different from that listed previously)
Explanation of What the Test Measures

Lab Activity 9.1 *(cont.)*

Test Method Number and Name for Evaluation of Abrasion
Explanation of What the Test Measures

Test Method Number and Name for Evaluation of Abrasion (different from that listed previously)
Explanation of What the Test Measures

Lab Activity 9.1 *(cont.)*

Test Method Number and Name for Evaluation of Durability
Explanation of What the Test Measures

Test Method Number and Name for Evaluation of Durability (different from that listed previously)
Explanation of What the Test Measures

Lab Activity 9.1 *(cont.)*

Test Method Number and Name for Evaluation of Safety (if applicable). See Chapter 10 of text.
Explanation of What the Test Measures

Test Method Number and Name for Evaluation of Safety (if applicable and different that listed previously)
Explanation of What the Test Measures

Lab Activity 9.2: PIECE GOODS APPEARANCE AND PERFORMANCE TESTING

Mount the tested specimens on cardstock and attach them to this worksheet.

Your Name

Name(s) of Teammate(s)

Before beginning any tests, record the temperature and relative humidity in your lab (if available).

Fabric Name	Basic Weave Structure	Fiber Content

If the test methods you are using are different from the following methods or if additional tests are conducted, use the empty grid to write in the standards you are using and the data collected.

Test Method	Name of Test	Report	Results	Results	°F °C %RH
AATCC 8	Colorfastness to Crocking: Crockmeter Method	Color staining	____ Dry rating	____ Wet rating	
Test Method	**Name of Test**	**Report**	**Results**	**Results**	**°F °C %RH**
ASTM D5034	Standard Test Method for Breaking Strength and Elongation of Textile Fabrics (Grab Test)	Breaking force—report breaking force to nearest 0.01 lb of force or 0.01 mN	Breaking force specimens	Calculate breaking force mean to nearest 0.01 lb of force or 0.01 mN ____ Mean lbf or mN	

Lab Activity 9.2 *(cont.)*

Test Method	Name of Test	Report	Results	Results	°F °C %RH
ASTM D2261	Standard Test Method for Tearing Strength of Fabrics by the Tongue (Single Rip) Procedure (Constant-Rate-of-Extension Tensile Testing Machine)	Tearing force—report tearing force to nearest 0.01 lb of force or 0.01 mN	Warp breaking force specimens 1. _____ 2. _____ 3. _____ 4. _____ 5. _____ _____ Warp mean lbf or mN	Filling breaking force specimens 1. _____ 2. _____ 3. _____ 4. _____ 5. _____ _____ Filling mean lbf or mN	
Test Method	**Name of Test**	**Report**	**Results**	**Results**	**°F °C %RH**
ASTM D3884	Standard Test Method for Abrasion Resistance of Textile Fabrics (Rotary Platform, Double-Head Method)	Abrasive wheel used Load in grams _____ Vacuum suction level _____	Abrasion specimens 1. _____ 2. _____ 3. _____	_____ Average cycles to failure	
Test Method	**Name of Test**	**Report**	**Results**	**Results**	**°F °C %RH**

Lab Activity 9.2 *(cont.)*

Test Method	Name of Test	Report	Results	Results	°F °C %RH
Test Method	**Name of Test**	**Report**	**Results**	**Results**	**°F °C %RH**
Test Method	**Name of Test**	**Report**	**Results**	**Results**	**°F °C %RH**

Lab Activity 9.3: COMPARISON GARMENT APPEARANCE AND PERFORMANCE TESTING

Mount the tested specimens on cardstock and attach them to this worksheet.

Your Name

Name(s) of Teammate(s)

Before beginning any tests, record the temperature and relative humidity in your lab (if available).

Garment Description	Brand

If the test methods you are using are different from the following methods or if additional tests are conducted, use the empty grid to write in the standards you are using and the data collected.

Test Method	Name of Test	Report	Results	Results	°F °C %RH
AATCC 8	Colorfastness to Crocking: Crockmeter Method	Color staining	______ Dry Rating	______ Wet Rating	
Test Method	**Name of Test**	**Report**	**Results**	**Results**	**°F °C %RH**
ASTM D3512 /D3512M	Standard Test Method for Pilling Resistance and Other Related Surface Changes of Textile Fabrics: Random Tumble Pilling Tester	Pilling rating	Specimen ratings 1. ______ 2. ______ 3. ______	______ Mean rating	

Lab Activity 9.3 *(cont.)*

Test Method	Name of Test	Report	Results	Results	°F °C %RH
ASTM D3787	Standard Test Method for Bursting Strength of Textiles—Constant-Rate-of-Traverse (CRT) Ball Burst Test	Bursting force at break. Report bursting force to nearest .01 lbs. of force or .05 N	Specimen Breaking Force 1. _____ 2. _____ 3. _____ 4. _____ 5. _____	Calculate bursting force mean to nearest .01 lbs. of force or .05 N _____ Mean lbf or N	

Test Method	Name of Test	Report	Results	Results	°F °C %RH
ASTM D3512 /D3512M	Standard Test Method for Pilling Resistance and Other Related Surface Changes of Textile Fabrics: Random Tumble Pilling Tester	Pilling rating	Specimen ratings 1. _____ 2. _____ 3. _____	_____ Mean rating	

Test Method	Name of Test	Report	Results	Results	°F °C %RH

Lab Activity 9.3 *(cont.)*

Test Method	Name of Test	Report	Results	Results	°F °C %RH

Test Method	Name of Test	Report	Results	Results	°F °C %RH

Test Method	Name of Test	Report	Results	Results	°F °C %RH

Lab Activity 9.3 *(cont.)*

Test Method	Name of Test	Report	Results	Results	°F °C %RH

Lab Activity 9.4: COMPARISON PROJECT GARMENT RESULTS AND PERFORMANCE SPECIFICATIONS

Your Name

Name(s) of Teammate(s)

Garment Description	Brand

Check one of the boxes below to indicate the specification you are using or write one in.

Specification Method	Name of Specification
☐ASTM D4154	Performance Specification for Men's and Boys' Knitted and Woven Beachwear and Sports Shirt Fabrics
☐ASTM D4156	Performance Specification for Women's and Girls' Knitted Sportswear Fabrics
☐	

If the test methods you used are different from the following methods or if additional tests were conducted, use the empty grid to write in the standards used and the mean data or ratings collected.

Test Method	Name of Test	Results	Results	Specification Max, Min, or Class/Rating	Pass or Fail
AATCC 8	Colorfastness to Crocking: Crockmeter Method	_____ Dry rating	_____ Wet rating		
ASTM D3787	Standard Test Method for Bursting Strength of Textiles—Constant-Rate-of-Traverse (CRT) Ball Burst Test	_____ Mean lbf or N			

Lab Activity 9.4 *(cont.)*

Test Method	Name of Test	Results	Results	Specification Max, Min, or Class/Rating	Pass or Fail
ASTM D3512/ D3512M	Standard Test Method for Pilling Resistance and Other Related Surface Changes of Textile Fabrics: Random Tumble Pilling Tester	______ Mean rating			
AATCC 150	Dimensional Changes of Garments after Home Laundering	______%DC length mean after 5 launderings	______%DC width mean after 5 launderings		
AATCC 61	Colorfastness to Laundering: Shade Change	______ Shade change mean rating			

Lab Activity 9.4 *(cont.)*

Test Method	Name of Test	Results	Results	Specification Max, Min, or Class/Rating	Pass or Fail

Mount before and after photos of comparison garment here.

Original Condition

Insert Photo "Before" Here

After Fifth Wear and Refurbishment

Insert Photo "After" Here

Lab Activity 9.4 *(cont.)*

ASTM D4154 and D4156 standard performance specifications do not contain minimum ratings for acceptability for pilling. Pilling is an important aspect of appearance. In your opinion as a consumer, do you find the pilling rating obtained for your garment acceptable or unacceptable? ☐Acceptable ☐Unacceptable Explain your answer.
Are there any tests results that failed to meet the performance specifications that you believe a consumer would be okay with? ☐Yes ☐No Explain your answer.

Lab Activity 9.4 *(cont.)*

Rank each brand your group tested (first place being the best and fifth place being the worst) based on overall appearance and performance.	
1.	
2.	
3.	
4.	
5.	

Rank each brand your group tested (first place being the best and fifth place being the worst) based on price and value.	
1.	
2.	
3.	
4.	
5.	

Were your brand rankings the same for both overall appearance and performance and price and value? ☐Yes ☐No Explain your answer.

NOTES

CHAPTER 10

Quality Assurance along the Supply Chain Lab

LAB OBJECTIVES:

- To inspect apparel items for fabric and garment defects.

Chapter 13 of *Apparel Quality: A Guide to Evaluating Sewn Products* focuses on the importance of quality control tools and inspection of apparel products. Companies maintain consistent quality by carefully monitoring, testing, and inspecting raw materials and garments from selection through production. When companies focus on producing quality garments, they have higher rates of customer satisfaction and brand loyalty.

There are three lab activities (worksheets 10.1, 10.2, 10.3) for this chapter, which focus on inspection of garments and completion of the customer satisfaction survey for the Comparison Project. Fabric and garment flaws will be identified in randomly selected garments from your wardrobe and the Comparison Project control garments. Additionally, you will work on completing Lab Activities 9.2 and 9.3, which may take several lab sessions to complete all of the testing. After you have complete Lab Activity 9.3, you can go on to Lab Activity 9.4.

PREPARATION AND SUPPLIES FOR CHAPTER 10 LAB ACTIVITIES

In preparation for Chapter 10 lab activities, each group member (teams of 2 to 3) will randomly select a garment from their wardrobe and bring it to lab for inspection. Everyone will bring their comparison project control garment. To complete Chapter 10 lab activities, you will also need the *Apparel Quality: A Guide to Evaluating Sewn Products*. In addition, continue working on completing testing of piece goods and comparison garment lab activities from Chapter 9. Be sure to bring all of the supplies and test specimens to lab.

Lab Activity 10.1: Inspection of Randomly Selected Garments

For Lab Activity 10.1: Inspection of Randomly Selected Garments, work in teams of two to three people. Each team member will randomly select one woven garment from his or her wardrobe and bring it to lab. As a team, you will inspect the garments for fabric and garment defects outlined in Chapter 13 of *Apparel Quality: A Guide to Evaluating Sewn Products*. Identify the defects contained in each of the garments. Tally how many defects each garment has and note the position/location of the defect within the garment. As a group, determine whether a consumer would find these defects acceptable or unacceptable, and explain why. Each group member will complete worksheet 10.1 for his or her own garment.

Lab Activity 10.2: Comparison Project Garment Inspection

For Lab Activity 10.2, work in your Comparison Project teams of four to five people. Each team member will inspect his or her own Comparison Project control garment for fabric and garment defects outlined in Chapter 13 of *Apparel Quality: A Guide to Evaluating Sewn Products*. Identify each defect contained in the Comparison Project control garment. Now that you are more familiar with the types of defects that can be found in garments, inspect your comparison shirt, and record the position/location of each defect by zone. In the industry, garments are divided into zones for inspection. These zones are determined by companies and are included as part of their quality assurance inspection and procedures specifications because zone designations are not standardized within the industry. One company may divide a garment into two zones, whereas another firm may divide the same style of garment into three zones. For the purpose of this activity, the Comparison Project shirts will be divided into two zones.

- Zone 1—areas of the garment that are highly visible (i.e., front, back, neckline, placket, top portion of sleeve, shoulder seam)
- Zone 2—areas of the garment that are less visible or hidden during everyday use (i.e., side seams, under arm seams, bottom hem)

In addition, you will rate each defect as critical, major, or minor.

- Critical—defects that can harm or pose a safety risk to the customer (i.e., broken needle or straight pin sewn into the garment)
- Major—defects that adversely affect the appearance, performance, or fit of the garment and may deter a customer from purchasing the item or cause them to return it
- Minor—defects that may not be easily detected by the customer at the time of purchase and will not prompt a customer to return the garment. These are visible, not structural, and do not impact fit of the garment.

Tally how many critical, major, and minor defects your garment has. As a consumer, determine if you find these defects acceptable or unacceptable, and explain why. Each group member will complete worksheet 10.2 for his or her Comparison Project garment.

Lab Activity 10.3: Comparison Garment Customer Satisfaction Survey

For Lab Activity 1.4, your group completed the Customer Expectation Survey. Now for Lab Activity 10.3: Comparison Garment Customer Satisfaction Survey, independently evaluate your satisfaction level with your Comparison Project garment. Keep in mind the expectations your team had for this style garment at the price classification selected.

Lab Activity 10.1: INSPECTION OF RANDOMLY SELECTED GARMENTS

Each team member, in collaboration with the rest of the team, will complete this worksheet for his or her woven garment. Document each fabric and garment defect present.

Your Name

Name(s) of Teammate(s)

Garment Description	Brand

Material Defects

Defects in Woven Fabrics	Number of Occurrences and Position/Location	Defects in Woven Fabrics	Number of Occurrences and Position/Location
Broken pick		Hole	
Coarse yarn/thick yarn		Jerk-in	
Double pick		Slub	
End out		Skew	
Fine yarn/thin yarn		Soiled yarn	
Float		Thin place	
Defects in Woven Fabrics	**Number of Occurrences and Position/Location**	**Defects in Woven Fabrics**	**Number of Occurrences and Position/Location**
Bow		Misregister	
Color out		Scrimp	
Color smear		Shaded	
Crease mark		Water spots	
Pin holes			

Subtotal the number of material defects (including the number of occurrences).

Subtotal of Material Defects	

Lab Activity 10.1 *(cont.)*

Garment Defects

Cutting Defects	**Number of Occurrences and Position/Location**	**Defects in Woven Fabrics**	**Number of Occurrences and Position/Location**
Bow		Incorrect ply tension	
Drill hole		Misaligned plies	
Deep notch		Mismatched fabric design	
Frayed edges		Wrong grain	
Fused edges			
Fusing Defects	**Number of Occurrences and Position/Location**	**Defects in Woven Fabrics**	**Number of Occurrences and Position/Location**
Blister/bubbling		Overfusing	
Delamination		Strike back/back-bleed	
Moiré effect		Strike through/bleed-through	
Construction and Sewing Defects	**Number of Occurrences and Position/Location**	**Defects in Woven Fabrics**	**Number of Occurrences and Position/Location**
Broken stitch		Seam repair	
Burst seam		Shading	
Cracked seam/ cracked stitches		Size defects	
Faulty zipper		Skipped stitch	
Incorrect SPI or SPC		Soil	
Irregular stitching		Thread discoloration	
Loose buttons		Twist garment	
Loose thread end		Unbalanced stitch tension	
Missing component		Uneven seams	
Needle damage		Unraveling buttonhole	
Open seam		Unrelated seam	
Raw edges		Visible stay stitch	
Ropy hem		Wavy seam	

Lab Activity 10.1 *(cont.)*

Construction and Sewing Defects	Number of Occurrences and Position/Location	Defects in Woven Fabrics	Number of Occurrences and Position/Location
Run-off / Overrun stitching		Wrong thread color	
Seam grin		Wrong thread type and size	
Seam pucker			

Subtotal the number of garment defects (including the number of occurrences).

Subtotal of Garment Defects	

Total number of Defects (material and garment)	

How does the group feel about the number of defects found in your garment?

Lab Activity 10.1 *(cont.)*

Does the group think the types of defects and locations found in your garment are acceptable or unacceptable? Explain.

Are there any defects that could have been prevented? ☐Yes ☐No List each defect that could have been prevented or repaired and explain how.

Lab Activity 10.2: COMPARISON PROJECT GARMENT INSPECTION

Each team member, in collaboration with the rest of the team, will complete this worksheet for his or her comparison garment. Document each fabric and garment defect present.

Your Name

Name(s) of Teammate(s)

Garment Description	Brand

Material Defects

Defects in Woven Fabrics	Zone (1 or 2)	Number of Critical Defects	Number of Major Defects	Number of Minor Defects
Barre				
Birdseye				
Hole				
Ladder/run				
Needle line				
Slub				
Skew				
Fabric Finishing Defects	**Zone (1 or 2)**	**Number of Critical Defects**	**Number of Major Defects**	**Number of Minor Defects**
Bow				
Color out				
Crease mark				
Pin holes				
Shaded				
Water spots				

Lab Activity 10.2 *(cont.)*

Subtotal the number of material defects by rating (including the number of occurrences).

Subtotal of Material Defects	Critical		Major		Minor	

Garment Defects

Cutting Defects	Zone (1 or 2)	Number of Critical Defects	Number of Major Defects	Number of Minor Defects
Bow				
Frayed edges				
Incorrect ply tension				
Misaligned plies				
Wrong grain				
Construction and Sewing Defects	**Zone (1 or 2)**	**Number of Critical Defects**	**Number of Major Defects**	**Number of Minor Defects**
Broken stitch				
Burst seam				
Cracked seam/cracked stitches				
Faulty zipper				
Incorrect SPI or SPC				
Irregular stitching				
Loose buttons				
Loose thread end				
Missing component				
Needle damage				
Open seam				
Raw edges				
Ropy hem				
Seam repair				
Shading				
Size defects				

Lab Activity 10.2 *(cont.)*

Construction and Sewing Defects	Zone (1 or 2)	Number of Critical Defects	Number of Major Defects	Number of Minor Defects
Skipped stitch				
Soil				
Thread discoloration				
Twist garment				
Unbalanced stitch tension				
Uneven seams				
Unraveling buttonhole				
Unrelated seam				
Visible stay stitch				
Wavy seam				
Run-off/overrun stitching				
Seam grin				
Seam pucker				
Wrong thread color				
Wrong thread type and size				

Subtotal the number of garment defects by rating (including the number of occurrences).

Subtotal of Garment Defects	Critical		Major		Minor	

Total the number of defects by rating.

Total Number of Defects (material and garment)	Critical		Major		Minor	

Lab Activity 10.2 *(cont.)*

Did you find any critical defects in this garment? ☐Yes ☐No If yes, describe the critical defect.

How do you feel about the total number and types of major defects found in this garment?

Lab Activity 10.2 *(cont.)*

Do you think the number of major defects found in this garment is acceptable or unacceptable? Explain.

How do you feel about the total number and types of minor defects found in this garment?

Lab Activity 10.2 *(cont.)*

Do you think the number of minor defects found in this garment is acceptable or unacceptable? Explain.

Are there any defects that could have been prevented? ☐Yes ☐No List each defect that could have been prevented or repaired and explain how.

Lab Activity 10.3: COMPARISON GARMENT CUSTOMER SATISFACTION SURVEY

Each team member will independently complete this survey for his or her Comparison Project garment.

Your Name

Name(s) of Teammate(s)

Garment Description	Brand

Rank the importance of each item on a scale of 1–5.
5 = Extremely Important *4 = Very Important* *3 = Moderately Important*
2 = Slightly Important *1 = Not Important*

Aesthetic Characteristics	**5**	**4**	**3**	**2**	**1**
Overall attractiveness of the materials, styling, and design of the garment for its intended use					
Maintained shape and appearance during wear					
Maintained shape and appearance after refurbishment (cleaning)					
Functional Characteristics	**5**	**4**	**3**	**2**	**1**
Ability of garment to perform for its intended use					
Comfort of the garment					
Fit of the garment					
Accuracy of sizing/true to size					
Durability for intended use					
Color maintained after cleaning; color did not rub off or bleed onto other items					
Ease of care					
External Factors	**5**	**4**	**3**	**2**	**1**
Price					
Value					
Brand					

Lab Activity 10.3 *(cont.)*

Were you satisfied with the fiber content of your garment? ☐Yes ☐No Explain your answer.

Will you continue to wear this garment? ☐Yes ☐No Explain your answer.

Lab Activity 10.3 *(cont.)*

How long do you expect this garment to last (useful life of the product)?

Were any of your aesthetic expectations not met? ☐Yes ☐No Explain your answer.

Lab Activity 10.3 *(cont.)*

Were any of your performance expectations not met? ☐Yes ☐No Explain your answer.

Were any of your price and brand expectations not met? ☐Yes ☐No Explain your answer.

Lab Activity 10.3 *(cont.)*

Do you believe this product was a good value for the price paid? ☐Yes ☐No Explain your answer.

Would you purchase this brand of product again? ☐Yes ☐No Explain your answer.

NOTES

GLOSSARY

(*Note*: All chapter references are to *Apparel Quality: A Guide to Evaluating Sewn Products*.)

100 Class Stitches. A category of chain stitches designated by ISO and ASTM in which stitches are formed by the intralooping or interloping of one needle thread passing through the fabric and held in place by subsequent loops. (Chapter 6)

100 Percent Inspection. Each individual garment within a lot is inspected, and acceptance or rejection is based on only the unit inspected. (Chapter 13)

10-Point Grading System. Standardized method used for inspecting piece goods that assigns penalty points to fabric flaws on a scale of 1 to 10 to determine if the roll will be accepted or rejected. (Chapter 13)

200 Class Stiches. A category of hand stitches designated by ISO and ASTM in which stitches are formed by hand or are machine simulations of hand stitches. (Chapter 6)

300 Class Stiches. A category of lockstitches designated by ISO and ASTM in which stitches are formed with a bobbin thread and one or more needle threads that pass through the fabric and interlace to secure each stitch. (Chapter 6)

400 Class Stitches. A category of chain stitches designated by ISO and ASTM in which stitches are formed with a looper thread and one or more needle threads that pass through the fabric and interlace with loops that interloop on the underside of the material. (Chapter 6)

4-Point Grading System. Standardized method used for inspecting piece goods that assigns penalty points to fabric flaws on a scale of 1 to 4 to determine if the roll will be accepted or rejected. (Chapter 13)

500 Class Stitches. A category of overedge stitches designated by ISO and ASTM in which stitches are formed by one group of threads penetrating through the fabric and held in place by intralooping (ISO term) or interloping (ASTM term) to cover the edge of the material where subsequent loops pass through the fabric to form the stitch. The threads can also be secured by one group of threads interlooping with loops formed by one or more interlooped groups before subsequent loops from the first group are passed back through the fabric to cover the edge and form the stitch. (Chapter 6)

600 Class Stitches. A category of cover stitches designated by ISO and ASTM in which stitches are formed by two or more groups of threads that enclose the raw edges of both surfaces of the fabric plies by covering them. The threads are cast on the surface of the fabric and then interlooped with loops of thread formed on the underside of the fabric. (Chapter 6)

AATCC (American Association of Textile Chemists and Colorists). A nonprofit organization that develops and publishes voluntary standardized test methods for international use. (Chapter 1)

Abrasion. Rubbing of a material against itself or another surface. (Chapter 12)

Acceptance Quality Limit (AQL). The total amount of defects allowed for an inspection sample to be accepted. (Chapter 13)

Acceptance Testing. Evaluation conducted to determine if a material or product meets the specified criteria for approval. (Chapter 11)

Aesthetic Characteristics. The overall attractiveness of the materials, styling, and design of the garment in relation to its intended use. (Chapter 1)

Aesthetics. The appearance, comfort, sound, and smell of a garment; any portion of a garment that engages the senses. (Chapter 1)

Agile Manufacturing. Manufacturing with focus on speed and flexibility to decrease lead time and meet market demands for continuous style change. (Chapter 8)

Air-Entangled Thread. A twisted thread that is formed when continuous filament fibers are passed through high-pressure air jets that cause them to become entwined. (Chapter 6)

Air Temperature. The degree of coolness or hotness of air. (Chapter 11)

Alcohol Ethoxylate. The primary active ingredient (surfactant) in wetcleaning detergent. (Chapter 12)

All-in-One Facing (Combination Facing). A facing that accommodates more than one garment area (i.e., neckline and armhole). (Chapter 4)

Amorphous Regions. The unorganized regions of a fiber. (Chapter 11)

Analyzer. Second or additional polarizer. (Chapter 11)

ANSI (American National Standards Institute). A private nonprofit organization that administers and coordinates the voluntary standardization system in the United States. (Chapter 1)

Anthropometric Data. Information gathered from measuring human populations to determine variations and commonalities in body dimension and shape for use in standardizing sizes of apparel and accessories. (Chapter 5)

Apparatus. Machine or equipment used for conducting a test. (Chapter 11)

Appearance. The overall visual aesthetic of a textile or garment. (Chapter 12)

Appearance Retention. Ability of a material or garment to maintain its aesthetic look during use, refurbishment, and storage. (Chapter 12)

Approved Suppliers. Vendors selected by a company to manufacture products based on reputation and negotiated quality level and price. (Chapter 13)

Aqueous Solution. Water and detergent mixture that acts as the solvent to dissolve water-soluble soils during laundering and wetcleaning. (Chapter 12)

ASTM International. Organization, formerly known as the American Society of Testing and Materials, that develops and publishes voluntary standards for international use. (Chapter 1)

Asymmetric (Informal Balance). A garment that is not the same on both sides of the center; design details vary from one side to the opposite side of the garment. (Chapter 2)

Atom. The smallest unit of an element. (Chapter 11)

Azo Dyes. A group of synthetic dyestuffs based on nitrogen used in textile and leather products. (Chapter 10)

Baffles. Paddles found inside a horizontal-axis washing machine, drycleaning machine, or dryer to keep garments moving by means of tumbling action. (Chapter 12)

Balance. When the right and left side of the garment appear to be even. (Chapter 2)

Balanced Plain Weave. Woven fabric constructed having the same size, type, and number of yarns in both the warp and filling directions. (Chapter 3)

Balanced Thread Tension. When the stitching treads interlock at the mid-point of the fabric layers providing a smooth, flat appearance. (Chapter 6)

Balanced Twist. The single yarns within a ply yarn typically possess an S twist, whereas the overall yarn containing the plies is twisted in the Z direction. Each of the yarns contains a specified number of turns per inch, per centimeter, or per meter to balance the twist. A ply yarn having balanced twist will not kink or twist back on itself when folded in half. (Chapter 6)

Ballast. Pieces of white or light color plain weave fabric that are added to a wash load during testing to maintain the specified weight required for the load. (Chapter 12)

Band Hem. An edge finish that utilizes a strip of material that is stitched to conceal the fabric raw edge along the hem. (Chapter 4)

Band Placket. Two finished strips of fabric that are used to form a lapped closure and finish the raw edge. (Chapter 4)

Barre. A knit fabric defect that gives the visual appearance of streaks across the width of the fabric. (Chapter 13)

Base Burn or Base Fabric Ignition. When a fabric with a raised surface is ignited and continues to burn the foundation yarns of the material. (Chapter 10)

Better Price Point. The price category that includes above-average prices for mass-produced apparel products with consumer expectations of better quality fabrics and more advanced construction details. (Chapter 1)

Bias-Faced Waistline. A facing cut on the bias that is used to contour and shape the waistline area. (Chapter 4)

Bias Facing. A narrow strip of fabric cut on the bias that is stitched along the opening to finish and stabilize the edge. (Chapter 4)

Bias Grain. A diagonal line that usually runs at a 45-degree angle from the perpendicular lengthwise and crosswise grainlines. (Chapters 2 and 4)

Bill of Materials (BOM). Part of a technical package that details component material specifications such as interlinings, trim, and findings for the production of a garment. (Chapter 5)

Birdseye Defect. The occurrence in a knitted fabric of two consecutive tuck stitches that are not intended. (Chapter 13)

Birefringence or Double Refraction. Direction and transmittance of polarized light used to determine the reflective index of a fiber's optical property. (Chapter 11)

Blind Hem. A hem finish where the raw edges are folded or covered with seam tape and sewn with a blind stitch so the stitches are not visible on the face of the garment. (Chapter 4)

Blind Tucks. A series of repeating parallel folds that are designed to have the fold meet at the stitch line of the next tuck. (Chapter 4)

Blistering or Bubbling. Visible pebbled appearance on the surface of the garment due to improper fusing. (Chapter 13)

Block. A tag board or digital pattern developed to a brand's specifications for a sample size that is used to develop first patterns of new designs or modifications of existing styles. (Chapter 2)

Bound Buttonhole (Welt Bound Buttonhole). Two separate strips of fabric are stitched to form the lips where the button will pass through. (Chapter 4)

Bound Hem (Welt-Finished Edge or Hong Kong Finish). A bias strip of material used to cover the raw edges of the material at the hemline of a garment. (Chapter 4)

Bound Placket. Two bindings are used to cover the raw edges of the opening without creating a lap. (Chapter 4)

Bow. The visual appearance of arcs across the width of woven or knitted fabric resulting from displaced yarns. (Chapter 13)

Brand. The reputation of a product or company conveyed through brand image, word mark, logo, product design, quality, marketing and promotion, distribution of goods, and customer service. (Chapter 1)

Break. Folds in the pant legs of trousers near the hem where it hits the instep of the foot. (Chapter 5)

Breaking Strength. Amount of force applied to a material before the yarns break and a tear occurs. (Chapter 12)

Bridge Price Point. The price category for mass-produced apparel products priced between the better and designer price categories with customer perceptions and expectations of high-quality fabrics and elements of designer apparel at a price point lower than designer products. (Chapter 1)

Broken Pick. A severed filling yarn in a fabric. (Chapter 13)

Broken Stitch. A severed stitch thread in a garment. (Chapter 13)

BS Bound Seam Class. ASTM International designation for assembling seams constructed with the raw edges of the seam allowance of one or more plies of fabric covered with a binding and stitched with one or more rows of stitches. (Chapter 7)

BSI (British Standards Institution). A nonprofit organization that develops and publishes standards used in the United Kingdom. (Chapter 1)

Budget Price Point. The lowest price category for mass-produced apparel products, geared toward a wide range of market segments. (Chapter 1)

Builder. Detergent additive that softens and controls the pH level of the water while protecting washing machines from corrosion. (Chapter 12)

Bundle Coupons. Used to track and compensate sewing operators for completed operations. (Chapter 8)

Bundle Ticket. A master list of operations for a bundle that includes bundle coupons, routing information, piece rate, style number, size, and shade number. (Chapter 8)

Buoyant Force. The force applied from beneath a submerged specimen by a lower density liquid than the specimen. (Chapter 11)

Burn Time or Flame Spread Time. The time measured on the flammability testing apparatus from the point of ignition to when the stop thread is severed through. (Chapter 10)

Burst Seam. Broken stitches within a seam due to improperly joined materials or tension on the garment. (Chapter 13)

Bursting Strength. Amount of multidirectional force applied to a material to cause the yarns to rupture, creating a hole in the fabric. (Chapter 12)

Button. A form of closure or decoration made from a variety of materials (plastics, wood, metals, natural products) in various shapes and sizes and attached to a garment by sewing through the center of the button or a button shank, or is secured by a prong. (Chapter 4)

Button Loop. A chain of thread, braid, cord, or fabric that is folded to form a U-shape and stitched into the seam to accommodate a button to pass through and remain secured. (Chapter 4)

Buttonhole. A finished opening made to accommodate a button to pass through and remain secured. (Chapter 4)

CA Number. A number issued by the Competition Bureau for the purpose of identifying Canadian manufacturers, importers, or distributers of apparel products. (Chapter 9)

Calculation. The formula used for computing test results data. (Chapter 11)

Calibration. Process of controlling the output from equipment or a testing standard to ensure accuracy of performance and results. (Chapter 11)

Care Labeling of Textile Wearing Apparel and Certain Piece Goods. U.S. labeling regulation 16 CFR 423 for disclosure of care instructions on garment labels. (Chapter 9)

CAS. Chemical Abstract Service. (Chapter 10)

CAS Number. The registration designation assigned to individual chemical substances located in the CAS database to provide information specific to a wide range of substances. CAS numbers are assigned by the American Chemical Society. (Chapter 10)

Casing. Fabric that is folded over and stitched at the edge of the garment or a separate strip of material that is applied to the garment to accommodate a drawstring or elastic. (Chapter 4)

CEN (European Committee for Standardization). An international nonprofit organization developing voluntary national standards for European countries. (Chapter 1)

Center Zipper Insertion (Slot Zipper). The seam edges of the opening are abutted over the center of the zipper chain to conceal it. (Chapter 4)

Certificate of Compliance (COC). Document provided by a supplier that certifies that goods comply with all of the required standards and specifications. (Chapter 13)

Certified Supplier. An established vendor that has been approved to produce products that require minimal inspection due to their strong quality control program and consistent record for manufacturing products that meet or exceed required specifications. (Chapter 13)

Char Length. The area of a flammability test specimen that is burned, charred, or damaged after exposure to an ignition source. (Chapter 10)

Charged Solvent. Drycleaning solvent that contains detergent and moisture to aid in stain removal. (Chapter 12)

Chemical Bonding. A chemical reaction between the fabric of the garment and the adhesive used to glue the seam. (Chapter 7)

Chlorine Bleach. A bleach used to whiten garments. (Chapter 12)

Citric Acid. Disinfectant used in wetcleaning detergents. (Chapter 12)

Class 1, Normal Flammability. A U.S. distinction for fabric flammability of plain or raised fiber surface fabrics. Plain surface fabrics must have a minimum burn time of 3.5 seconds, and raised fiber surface fabrics must possess a minimum burn time of 7 seconds or cannot exceed a surface flash of 7 seconds without causing a base burn when tested. Class 1 fabrics can be sold in the United States. (Chapter 10)

Class 2, Intermediate Flammability. A U.S. distinction for fabric flammability for raised surface fabrics only. Raised fiber surface fabrics that possess a burn time between 4 and 7 seconds and result in a base burn. Class 2 fabrics can be sold in the United States. (Chapter 10)

Class 3, Rapid and Intense Burning. A U.S. distinction for fabric flammability. Highly flammable fabrics that possess a burn time less than 4 seconds and result in a base burn. Class 3 fabrics cannot be sold in the United States. (Chapter 10)

Class A Children's Nightwear. The EU classification system that designates flammability of children's nightwear by age group or garment type and provides requirements for minimum flammability standards. Children's nightwear (excluding pajamas) for children over 6 months old up to 14 years of age, having a height of 68 cm (26.77 inches) up to 176 cm (69.29 inches) for girls and up to 182 cm (71.65 inches) for boys. Class A exhibits a flame spread of less than 15 seconds, has a char length less than 520 millimeters (20.47 inches), does not breaking the third marker thread when tested, and cannot exhibit surface flash. (Chapter 10)

Class B Children's Nightwear. The EU classification system that designates flammability of children's nightwear by age group or garment type and provides requirements for minimum flammability standards. Children's pajamas for children over 6 months old up to 14 years of age, having a height of 68 cm (26.77 inches) up to 176 cm (69.29 inches) for girls and up to 182 cm (71.65 inches) for boys. Class B exhibits a flame spread of less than 10 seconds, has a char length less than 520 millimeters (20.47 inches), and does not breaking the third marker thread when tested. (Chapter 10)

Class C Children's Nightwear. The EU classification system that designates flammability of children's nightwear by age group or garment type and provides requirements for minimum flammability standards. Babies' nightwear produced for infants up to 6 months old and up to 68 cm (26.77 inches) in height. Class C products do not require testing; therefore, there are no minimum flammability requirements. (Chapter 10)

Closure. Finding used to secure a garment opening. (Chapter 4)

CNIS (China National Institute of Standardization). A nonprofit organization that develops standards to be implemented by the government within China. (Chapter 1)

Coarse Yarn or Thick Yarn. A yarn within a fabric that has a larger linear density than the other yarns it is adjacent to. (Chapter 13)

Color. Created through the absorption and reflectance of light. The seven color classifications in the visual spectrum include red, orange, yellow, green, blue, indigo, and violet. (Chapter 2)

Color Change. Loss or change in color. (Chapter 12)

Color Evaluation. Assessing color against standards throughout the design development and production processes. (Chapter 2)

Color Management. The monitoring of color throughout the development, manufacturing, and distribution processes. (Chapter 2)

Color Match. When the spectral data of a color sample and the color standard are the same or near identical. (Chapter 2)

Color Out. Missing color within a print design. (Chapter 13)

Color Permanence. Ability of dyes, pigments, and brighteners used for coloring textile materials to remain vibrant, stable, durable, and unchanged. (Chapter 12)

Color Production Standards. Approved reproducible hues that are consistent during manufacturing across a variety of materials and dye formulations. (Chapter 2)

Color Smear. Smudged color on a printed fabric. (Chapter 13)

Color Specifications. Requirements for matching color standards that include the dye or pigment formulations for reproducing color on designated materials in order to match color standards. (Chapter 2)

Color Staining. Transfer of color from one material to another. (Chapter 12)

Color Standards. Color samples by which product materials (fabrics, thread, findings, and trims) are used to match to ensure color consistency between materials. (Chapter 2)

Colorants. Dyes, pigments, and optical brighteners used to apply color to fibers, yarns, fabrics, and garments. (Chapter 3)

Colorfastness. Resistance to change color characteristics or transfer colorant(s) to adjacent materials. (Chapter 12)

Colorimeter (Spectrophotometer). An instrument used to digitally measure color in a numeric representation. (Chapter 2)

Colour Index. An online publication that provides the chemical formulations for dyes and pigments. (Chapter 2)

Compliance. The ability of a material, finding, or apparel product to conform to established standards and specifications. (Chapter 1)

Concave Darts. The legs of the dart curve toward the body from the fullest point converging to a diminishing point. (Chapter 4)

Conditioning. Exposure of specimens and materials to standard atmospheric conditions to prepare them for testing. (Chapter 11)

Conditioning Room, Chamber, or Cabinet. Contained place used to bring samples or specimens up to the designated standard atmospheric conditions for testing. (Chapter 11)

Conformance. The degree to which the design and performance of a product meets established standards. (Chapter 1)

Conformity Assessment. The process for monitoring whether products, materials, and processes are meeting required standards and specifications. (Chapter 1)

Construction Details Sheet. Part of a technical package that includes detailed information for assembling a garment for production. Product details such as trims and decorative stitching are also identified. (Chapter 5)

Contemporary Price Point. The price category that is between better and bridge and offers very trendy fashion-forward apparel items for the junior and misses markets. (Chapter 1)

Continuous Placket. One uninterrupted binding used to finish the edges. (Chapter 4)

Contoured Waistband. A shaped band and facing created to contour the waistline of the garment and finish the edge. (Chapter 4)

Control. Fabric sample or a garment used as a standard for drawing comparisons. (Chapter 11)

Control Measurements. A sizing system's body measurements that are used for designating the appropriate garment size for the customer. (Chapter 5)

Convex Darts. The legs of the dart curve away from the body from the fullest point converging to a diminishing point. (Chapter 4)

Core Spun Thread. A continuous filament fiber core that is wrapped with spun yarn and twisted. (Chapter 6)

Cost Per Wear. A calculation used to determine the value of a product by dividing the purchase price of a garment by the number of times it has been worn. (Chapter 1)

Cotton Count (cc or Ne_c). Linear density in length per unit mass of cotton and cotton-blended yarns measured in 840-yard lengths per pound. (Chapter 3)

Courses. Rows of knit stitches or horizontal loops that run in the crosswise direction of knitted fabrics. (Chapter 3)

Covalent Bond. The result of a chemical reaction that causes pairs of electrons to be shared between atoms. (Chapter 7)

Cracked Seam or Cracked Stitches. Broken stitch or seam in garments made from knitted fabrics that have been stretched beyond capacity. (Chapter 13)

Crease Mark. A fabric portion that has inadvertently been folded and set/pressed or not removed properly during finishing and cannot be pressed out. (Chapter 13)

Crease Retention. Ability to maintain a crease intentionally set into the fabric. (Chapter 12)

Critical Defects. Garment flaws having the potential to cause minor to fatal injuries or unsafe conditions during use and maintenance of the product. (Chapter 13)

Crocking. Color rub-off caused by loose dye on the surface of an item. (Chapter 12)

Cross-Grain. Runs in the filling or courses direction of a fabric that are perpendicular to the selvedge. (Chapter 4)

Cross Tucks. A series of parallel and perpendicular folds created in a grid pattern and stitched down. (Chapter 4)

Crystalline Regions. Organized regions of a fiber. (Chapter 11)

Cut-Make-Trim (CMT). Method of sourcing that uses vendors contracted to spread and cut fabric and then assemble the component parts (garment pieces, findings and trim) into garments according to specifications provided by the contracting firm. (Chapter 8)

Dart. A triangular fold created to take up excess fabric in a garment in an effort to shape a garment section around a body contour. (Chapter 4)

Dart Equivalent. Integrates the dart into a shaped seam to fit a body contour. (Chapter 4)

Dart Slash. One leg of the dart is gathered to shape around a body contour or add fullness. (Chapter 4)

Dart Tucks. Fabric that is folded back on itself to form a dart stitched to a designated length; the remaining fabric is released rather than converging into a point. (Chapter 4)

DC Inspections. Evaluations of apparel product shipments that are conducted at distribution centers. (Chapter 13)

Decitex (dtex). Mass per unit length of fine filament yarns from silk or manufactured fibers measured in grams per 10,000 meters. (Chapter 3)

Decorative Dart. The triangular folded portion of a dart that is formed/exposed on the exterior of the garment. (Chapter 4)

Deep Notch. A visible hole in the garment due to a notch that was cut too long. (Chapter 13)

Defect. A flaw in material or sewn product that is not acceptable for the planned quality level. (Chapter 13)

Defective. A product having one or more flaws that prevent it from satisfying the appearance or performance functions for its intended use. (Chapter 13)

Delamination. The breakdown of the adhesive bond between an interlining and a shell fabric. (Chapter 13)

Demographic Data. Statistical information about a population that includes age, gender, income, education, geographic area, family size, housing type, nationality/ethnicity/race, marital status, occupation, spending patterns, and religious affiliations. (Chapter 1)

Demonstrated Capacity. The total quantity of goods produced at a particular quality level within a specific amount of time. (Chapter 8)

Denier (den). Mass per unit length of filament yarns from silk or manufactured fibers measured in grams per 9,000 meters. (Chapter 3)

Design Detail. An element of a garment that creates visual interest within a silhouette. (Chapter 2)

Design Ease (Style Ease). Functional ease plus any additional fullness required to create the desired style or silhouette. (Chapters 4 and 5)

Design Specifications. Standards for styling details, design features, and characteristics of an apparel item in relation to its aesthetic appeal. (Chapter 1)

Designer Price Point. The highest price category for mass-produced apparel products with customer expectations of exclusive design, high-quality fabrics, and construction details for a narrowly defined niche target market. (Chapter 1)

Destructive Tests. Testing directly on a sample or specimen that must be cut or damaged in order for testing and evaluation to occur. (Chapter 11)

Digital Print. Fabric patterns applied to the surface of fabrics with a digital inkjet printer containing disperse dyes or pigment inks. (Chapter 3)

Dimensional Change. Change in the physical length or width of a material or apparel; shrinkage or growth. (Chapter 12)

Dimensional Pleats. Creased-ridge fabric folds that are permanently set into a pattern. (Chapter 4)

Dimensional Stability. Ability of textile materials and garments to retain their shape in the length and width directions when subjected to specific conditions relating to temperature and humidity. (Chapter 12)

Direct Yarn Number System. System for measuring the mass/weight per unit length of filament yarn. (Chapters 3 and 11)

Disodium Distyrylbiphenyl Disulfonate. Optical brightener. (Chapter 12)

Dobby Weave. Woven geometric patterned created during the weaving process. (Chapter 3)

Domestic Sourcing. Manufacturing sourced within the country in which the goods will be sold. (Chapter 8)

Donning and Doffing. The ease of putting a garment on and taking it off again. (Chapter 4)

Double Cloth. Two fabrics woven on the same loom where the two layers are interlocked by another set of yarns that is interlaced to attach the layers together. (Chapter 3)

Double-Ended Dart ("Fish-Eye" Dart or Double-Pointed Dart). Dart that originates at the apex level, extends to the hip level, and contours the fabric to the curves of the body from the bust to the waist to the hips. (Chapter 4)

Double-Folded Hem. A hem finish that is created by folding the materials edge two times then it is stitched. (Chapter 4)

Double Pick. Two filling yarns in a woven fabric located within the same shed. (Chapter 13)

Drag Lines. Horizontal or vertical folds in a garment that are not planned as part of the design, indicating fitting problems that need to be resolved. (Chapter 5)

Drill Hole Defect. A visible puncture in the fabric due to a misplaced drill mark. (Chapter 13)

Dry Finish. Chemical application in the form of a liquid or foam used to change the physical performance characteristics of the fabric. (Chapter 3)

Dry Prints. Dry pigment adhered to fabric by means of a resin binder that must be heat cured to form a pattern on the surface of a fabric. (Chapter 3)

Drycleaning. A method of refurbishment that uses a nonaqueous organic solvent for cleaning. (Chapter 12)

Dryer Drum. The chamber in which clothes are dried. (Chapter 12)

Dry-to-Dry Closed Loop Machines. Drycleaning machines where both the washing and drying takes place in one machine and the residual solvent vapors are recirculate to recover it. (Chapter 12)

Dry-to-Dry Vented Machines. Drycleaning machines where both the washing and drying takes place in one machine and the residual solvent vapors are vented outside the building. (Chapter 12)

Durability. Resistance to physical and mechanical deterioration. (Chapter 1)

Dyes. Complex water-soluble organic molecules used to add color to fibers, yarns, fabrics, or garments. (Chapter 3)

Ease. The amount of material in a garment designed to accommodate body movement. (Chapter 4)

EDANA (European Disposables and Nonwovens Association). A nonprofit organization that develops and publishes standards primarily for the nonwovens industries in Europe, the Middle East, and Africa. (Chapter 1)

Edge Abrasion. Caused by the edge of a garment rubbing against another surface. (Chapter 12)

EF Edge Finish Seam Class. ASTM International designation for assembling seams constructed with one or two plies of fabric used to cleanly finish the unfinished or raw edge of a garment. (Chapter 7)

Elasticized Waistline. Elastic applied directly to the waistline opening or inserted into a casing to provide stretch. (Chapter 4)

Element. The simplest chemical substance that is the basis for building molecular structures and cannot be further broken down using chemical means. (Chapter 11)

Emphasis. Focal point of a garment. (Chapter 2)

End-Item Inspection or Finished Product Inspection. An evaluation of garment quality that occurs once the production run is 100 percent assembled and a minimum of 80 percent is packaged. (Chapter 13)

End Out. A missing warp yarn within a fabric. (Chapter 13)

Engineered Print. A fabric print that is strategically placed and created to fit into a garment's shape/design. (Chapter 2)

Environmental Trend Research (Market and Global Trends Research). Gathering, evaluating, and tracking economic, political, cultural, and social trends and technology influences that have occurred over the past 12 months to provide insight for projections for the coming year(s) relating to factors that impact consumer spending. (Chapter 2)

Enzymes. A detergent or pretreat laundry product additive that aids in the removal of protein and starch-based soils. (Chapter 12)

Evaluation. Rating using photographic standards, visual standards, or gray scales. (Chapter 11)

Even Twill. A balanced twill weave having the same number of yarns passing over and under in the warp and filling directions. (Chapter 3)

Exposed Zipper Insertion. Zipper stitched into the garment to show the teeth and part of the zipper tape. (Chapter 4)

Extended Facing. The facing cut as part of the shell garment and folded back on itself to provide a soft draped edge. (Chapter 4)

Extracting. The process of spinning to remove excess moisture during the cleaning processes. (Chapter 12)

Extrinsic Quality Cues. External influences that contribute to the consumer's perception of quality such as price, image, and reputation of the brand and retailer, country of origin, advertising and marketing, and visual presentation. (Chapter 1)

Fabric. Substrate composed of fiber or fiber made into yarns that have been woven, knitted, or chemically, thermally, or mechanically bonded. (Chapter 3)

Fabric Construction. Structure of a material. (Chapter 3)

Fabric Count. The number of threads per inch or per 25 mm in both the warp and fill directions or the number of wale loops and course loops per inch or per 2.5 cm. (Chapters 3 and 11)

Fabric Failure. Force applied to a garment seam where the construction of the seam is stronger than the material of the product, causing the material to tear; the rupture of yarns within a fabric. (Chapters 7 and 12)

Fabric Fallout. Wasted fabric after garment parts have been cut. (Chapter 8)

Fabric Sheet. Part of a technical package that details fabric specification information for the production of a garment, such as fabric name, construction, weight, width, fiber content, color, finish (if applicable), and placement on the garment. (Chapter 5)

Fabric Specifications. Detailed information about fiber content, yarn composition, fabric structure, construction, weight, finishes, defect tolerances, color standards, and performance requirements for materials used in apparel production. (Chapter 5)

Fabric Thickness. The dimension between the top/face and bottom/back surfaces of a material. (Chapters 3 and 11)

Fabric Weight. Mass per unit area of a length of material. (Chapters 3 and 11)

Faced Hem. A piece of material (facing) shaped the same as the garment edge, stitched to the garment to conceal the raw edges and stabilize the hem. (Chapter 4)

Faced-Slashed Placket. Used to conceal the edges and provide stability to the opening when a vent or seam is not planned and no overlap is desired. (Chapter 4)

Facing. A piece of fabric that contours the edge of the outer portion of the garment to finish and stabilize the garment opening. (Chapter 4)

Factory Layout. Physical arrangement of the space within a manufacturing plant that contains areas for production, administration, raw materials storage, inspection and quality control, and employee service. (Chapter 8)

Faulty Zipper. A broken zipper or one that does not function properly. (Chapter 13)

Features. Physical characteristics or special components that enhance and support product performance. (Chapter 1)

Fiber. The smallest unit within the structure of a textile material. (Chapter 3)

Fiber Density. Mass per unit volume of a fiber. (Chapter 11)

Fiber Identification. Verification of fiber content. (Chapter 11)

Filament Fibers. Long continuous strands of fiber. (Chapter 3)

Findings. Any materials other than the shell fabric used to construct an apparel item such as support materials, shaping devices, trims, and surface embellishments. (Chapter 3)

Fine Yarn or Thin Yarn. A yarn within a fabric that has a smaller linear density than the other yarns it is adjacent to. (Chapter 13)

Finishing Specifications. Detailed information for completing the garment's final appearance, such as pressing, tagging, folding, or hanging. (Chapter 5)

First Pattern. The initial pattern developed for a sample garment that includes seam and hem allowances. (Chapter 2)

First Sample. The initial prototype constructed to provide a means for evaluating the garment for fit, function, and overall aesthetic appearance. (Chapter 5)

Fit. The relationship between the body and the size and styling of the garment. (Chapter 4)

Fit History. Documentation of changes made to measurements of the garment during the patternmaking and sampling process. (Chapter 5)

Flame Retardant. A chemical finish applied to give textiles flame resistance to prevent the spread of fire when a material is exposed to a flame. (Chapter 10)

Flammability. The ability of a material to ignite and burn. (Chapter 10)

Flange Dart. A pleat that is formed in the fabric, stitched down to a specified length, and then released at the opposite end to fit the contours of the body or add fullness to an area. (Chapter 4)

Flaring. When the horizontal balance of a garment is not level causing the garment to hang improperly. (Chapter 5)

Flat Abrasion. Caused by two plane surfaces rubbing together. (Chapter 12)

Flat Pleats. Folds of fabric creased, stitched to a desired length, or left unpressed that occur singly, in groups, or in an evenly spaced repeating pattern. (Chapter 4)

Flex Abrasion. Repeated folding that causes the yarns to wear against each other. (Chapter 12)

Float Defect. Yarn that passes over several other yarns rather than interlacing with each yarn. (Chapter 13)

Fluorinated Organic Compounds. Man-made chemical compounds used in apparel products to provide water, oil, or grease resistance. (Chapter 10)

Fly-Front Concealed Zipper Application. A lap formed when one side of a zipper is stitched to a facing that extends slightly beyond the closure; it can accommodate two fasteners such as a zipper and snaps. (Chapter 4)

Formaldehyde. A chemical composed of hydrogen, oxygen, and carbon used as a binder for dyes and pigments when combined with other compounds or applied as a finish for some wrinkle-free apparel items. (Chapter 10)

Frayed Edge Defect. Garment panel edges with displaced yarns. (Chapter 13)

French Dart. A diagonal dart that extends from the bust to the side seam at any point between the waist area and the hip. (Chapter 4)

FS Flat Seam Class. ASTM International designation for assembling seams constructed with the raw edges of the material plies abutted or slightly overlapped and joined together with stitching that covers the joint. (Chapter 7)

Full-Fashioned. Two-dimensional garment panels that are shaped as they are knitted and emerge from the knitting machine ready to assemble. (Chapter 4)

Functional Characteristics. The physical features of a garment related to fit, durability, effectiveness, and ease of care. (Chapter 1)

Functional Ease (Wearing Ease). The appropriate amount of fullness added to a garment to allow for body movement; fullness added to a garment pattern to allow for body movement. (Chapters 4 and 5)

Fur Products Labeling Act (FPLA). U.S. labeling regulation 16 CFR 301 for labeling garments containing animal fur. The regulation outlines the need to disclose the fiber content; registered identification number or identity of the manufacturer, or importer, distributor or retailer; and country of origin. (Chapter 9)

Fused Edges. Edges of a garment panel that have melted and bonded with other layers in the fabric spread during cutting. (Chapter 13)

Garment Opening. An area of a garment that allows access for the body to enter it. (Chapter 4)

Gathers. Even distribution of fullness in a garment that is created by drawing in one or more parallel rows of machine gathering stitches. (Chapter 4)

Gauge. The fineness of the knit; the number of needles per inch, per 2.5 centimeters, or per 25 millimeters used in a machine to create a knit fabrics; the size of the knit stitch. (Chapter 3)

Gauntlet Button. A button and buttonhole positioned in the sleeve placket to secure the opening and eliminate gaping. (Chapter 4)

Godet. Triangular or rounded panel that is inserted and sewn into a seam to provide fullness. (Chapter 4)

Golden Ratio (Divine Proportion). The irrational mathematical constant equal to approximately 1.618 that is used for aesthetically dividing or sectioning in design. (Chapter 2)

Gore. Vertical panels that are shaped to taper to the waist and are seamed together to provide fit. (Chapter 4)

Grade Rules. Directions for proportionately increasing and decreasing a garment pattern at specified points and designated amounts to produce all sizes in the range for mass production. (Chapter 5)

Grading. The measurements of the production pattern (created in the sample size) that are readjusted (increased or decreased) in designated areas in an effort to create all the sizes in the range. (Chapter 2)

Growth. An increase in the dimension of a garment after cleaning. (Chapter 12)

Gusset. Diamond-shaped inset stitched into garment areas to provide fullness where additional ease is needed. (Chapter 4)

Harmony. When all of the design elements of a garment work together to create a unified overall pleasing aesthetic. (Chapter 2)

Harness. Rectangular loom frames that hold the heddles threaded with yarns. (Chapter 3)

Heddles. Needle-like wires in the harness of a loom threaded with yarns. (Chapter 3)

Hem. A finish used to provide stability and cover fabric raw edges along the bottom edge of a garment or sleeves. (Chapter 4)

Hemline. The area of a garment where the hem is to be folded, faced, or finished. (Chapter 4)

Hemmed-Edge Placket. A band applied to finish the edges at the sleeve hem; a soft pleat is formed when folded back. (Chapter 4)

Hole. An unwanted opening in fabric due to damaged yarns. (Chapter 13)

Hook and Eye and Bar. A metal fastener that can be paired with an eye or bar to secure a garment closed. Hooks can also be paired with a fabric loop to form a closure. (Chapter 4)

Hue. The purest form of color. (Chapter 2)

Hydrogen Peroxide. The primary active ingredient in oxygen bleach. (Chapter 12)

Hydrophilic. Water-loving; having an affinity for water. (Chapter 12)

Hydrophobic. Water-repelling; having an aversion to water. (Chapter 12)

Hygrometer. A psychrometer having a wet-bulb thermometer that is kept moist through ventilation by aspiration and a dry-bulb thermometer. (Chapter 11)

Incorrect Ply Tension. Fabric layup that is spread too tight or too loose. (Chapter 13)

Incorrect SPI or SPC. The wrong number of stitches per inch or per centimeter. (Chapter 13)

INDA (Association of the Nonwoven Fabrics Industry). A nonprofit organization that develops and publishes international standards for the nonwoven materials industry. (Chapter 1)

Indirect Yarn Number System. System for measuring the length per unit of mass/weight of spun yarn made from staple fibers. (Chapter 11)

Infrared Spectrophotometer. Apparatus used for measuring the electromagnetic radiation to determine the chemical compounds within a specimen. (Chapter 11)

Infrared Spectroscopy (IR Spectroscopy). Method for chemically analyzing fiber specimens through measurement using infrared radiation. (Chapter 11)

Infrared Spectrum. The portion of wavelengths with the electromagnetic spectrum that emit infrared radiation. (Chapter 11)

In-plant Audits. Evaluation of a factory to monitor the management, efficiency, quality control, and working conditions of a manufacturing plant. (Chapter 13)

In-plant Inspections. Evaluations that occur in the factory where the products are produced. (Chapter 13)

In-process Inspection, In-line Inspection, or Dupro Inspection. Evaluation of garments during production. (Chapter 13)

Insoluble Compound Deposits. Greasy film or soap scum that results from a reaction between soap and hard water minerals. (Chapter 12)

Inspection. The process for evaluating factories in relation to capacity and quality control, function and appearance of materials and components, random selection of production garments to identify defects and deviations from contracted specifications to ensure that quality standards are being met; the process of testing and visually examining fabric, component parts, and sewn products at various stages of completion and checking garment dimensions to determine if they comply with required specification standards. (Chapters 1 and 13)

Inspection by Random Sampling. A method of inspection where a specific percentage of a shipment is inspected to determine acceptance or rejection of an entire shipment and is based solely on those garments evaluated. (Chapter 13)

Inspection by Spot Check Sampling. A method of inspection where products are arbitrarily selected for evaluation and decisions for acceptance or rejection that are based solely on the garments evaluated. (Chapter 13)

Inspection by Statistical Sampling. A method of inspection where a specific percentage of a shipment is inspected to determine acceptance or rejection of an entire shipment based on statistics. (Chapter 13)

Inspection Specifications. Standards developed by a company that are used to indicate the stages in which inspection will occur. (Chapter 13)

Inspectors. Individuals trained to conduct visual examinations of raw materials and sewn garments to determine if they will meet brand or customer expectations in relation to appearance, size, performance, and function. (Chapter 13)

Interlacing. A stitch formed by two or more sewing threads passing through fabric forming a loop that intertwines around a different loop of thread. (Chapter 6)

Interlooping. The stitch formation created with two or more threads passing through one or more loops formed by different threads. (Chapter 6)

Intralooping. A term used by ISO to indicate the stitch formation of a single sewing thread when it creates a loop that is then looped back through it. (Chapter 6)

Intrinsic Quality Cues. The physical features, performance characteristics, and product benefits consumers use to make determinations about quality. (Chapter 1)

Invisible Zipper Insertion. The chain of the zipper is completely concealed when closed and appears like a seam. (Chapter 4)

Irregular Stitching. Garment stitching that has not been sewn straight. (Chapter 13)

ISO (International Organization for Standardization). The largest organization for developing and publishing standards worldwide. (Chapter 1)

Jacquard Weave. Woven figured pattern or motif woven into the structure of the fabric. (Chapter 3)

Jerk-in. An extra filling yarn that has been inadvertently pulled into the fabric and does not extend across the width of the woven material. (Chapter 13)

JISC (Japanese Industrial Standards Committee). Japanese committee that develops and administers the JIS regulatory standards for use in Japan. (Chapter 1)

JSA (Japanese Standards Association). Publishes the JIS regulatory standards for use in Japan. (Chapter 1)

Key Words. Important terms used within a testing standard. (Chapter 11)

Kilotex (ktex). Mass per unit length of filament yarns of thick spun yarns from silk or manufactured fibers measured in kilograms per 1,000 meters. (Chapter 3)

Knit-and-Wear or Seamless Apparel. Three-dimensional preshaped garments that are knitted to fit the shape of the body and do not require any additional operations for assembly; they emerge from the knitting machine ready to wear. Some garments require minimal operations or finishing. (Chapter 4)

Lab Dip. Physical or digital representations of dyed fabric that are evaluated against color standards for accuracy of color matching. (Chapter 2)

Lab Samples. Materials or garments obtained from the lot sample to supply specimens for testing. (Chapter 11)

Label Sheet. Part of a technical package; details label specification information for the production of a garment. (Chapter 5)

Ladder or Run. A vertical line of unraveled stitches in a knitted fabric or garment. (Chapter 13)

Landed Duty Paid Supplier (LDP). A supplier responsible for the landed value of a product plus any import duties such as shipping, duty, delivery, insurance, and customs clearance costs. (Chapter 8)

Lap Zipper Insertion. The zipper is stitched to the back-folded opening while the other garment opening is folded and stitched along the zipper tape to form a tuck to conceal the zipper. (Chapter 4)

Laser Welding. Infrared laser technology transmitted through materials to melt a specified portion of the fabric's surface to bond it together. (Chapter 7)

Laundering. Cleaning process that utilizes aqueous detergent solution and mechanical action to remove soils, followed by rinsing, extracting, and drying to refurbish the garment. (Chapter 12)

Lauryl Trimethyl Ammonium Chloride. A conditioner added to wetcleaning detergent to prevent shrinkage and felting. (Chapter 12)

Layup. Total number of plies in a fabric spread. (Chapter 13)

Lead. A highly toxic naturally occurring chemical element found in the crust of the earth and used in some garment findings made from metal or plastic or in textiles treated with surface coatings or printing. (Chapter 10)

Lead Time. The amount of time between when an order is released to start production to when it is shipped. (Chapter 8)

Lean Manufacturing. Production of small quantities of merchandise close to the time needed. (Chapter 8)

Left-Hand Twill. Woven twill fabric having diagonal lines that extend from the lower right to the upper left. (Chapter 3)

Letter Code Sizing. Garment size designation reported by one or more letters such as XS or M. (Chapter 5)

Lifestyle Data. The social and psychological factors that motivate consumers to buy such as life stage, reference groups/peers, social class, personality, attitudes and values, generation group, and cultural preferences based on ethnic or cultural influences. (Chapter 1)

Line. A continuous stroke or contour used to create a silhouette or form shapes within a garment. (Chapter 2)

Linen Lea (lea). Linear density in length per unit mass of linen and linen-like yarns measured in 300-yard lengths per pound. (Chapter 3)

Linking (Looping). Attaching knit trim components to the body of the garment by matching each stitch and joining them together. (Chapter 4)

Logistics. Distribution process. (Chapter 8)

Loose Buttons. Buttons that have not been properly secured to the garment. (Chapter 13)

Loose-Fitting Sleepwear. A Canadian designation for children's night garments and sleepwear up to size 14X that is designed to fit looser around the body. (Chapter 10)

Loose Thread End. Thread end on a garment that has not been properly trimmed. (Chapter 13)

Lot Samples. Randomly selected garments or materials taken from one or more main stock or production lot shipments used for conducting acceptance testing. (Chapter 11)

LS Lapped Seam Class. ASTM International designation for assembling two or more plies of material that are overlapped with the raw edges exposed or the seam allowance folded under and joined with one or more rows of stitching. (Chapter 7)

Maintenance. Care and repair of a garment. (Chapter 1)

Major Defects. Flaws that impact the functional performance of a product; it will lead to failure at some point during usage. (Chapter 13)

Make-Through or Whole Garment Production System. A production process where one individual assembles a garment from start to finish. (Chapter 8)

Mandatory/Regulatory Standards. Test methods and specifications required by law that are enforced by government. (Chapter 1)

Manity Sizing. Vanity sizing used in the menswear industry. (Chapter 5)

Manufacturer-Defined Quality. The level of excellence of a product that is determined by its ability to conform to production standards and specifications. (Chapter 1)

Marker. The planned layout of pattern pieces used for cutting fabric for garment assembly. (Chapter 8)

Material Specifications. The designated performance expectations required for all materials used to complete a garment style. (Chapter 1)

Mean (x-bar) Average of the data obtained from a specified number of observations. (Chapter 11)

Overcut. Creating a garment pattern that is larger to accommodate the initial shrinkage that occurs during cleaning. (Chapter 12)

Overedge Hem. Stitching is formed over the fabric's raw edge to create a garment finish at the hem. (Chapter 4)

Overfusing. Weak bond between the interlining and shell materials due to excessive heat, time, or pressure during fusing, causing the resin to transfer completely into the garment fabric. (Chapter 13)

Oxidizing Bleach. Chlorine and oxygen bleach used for cleaning textiles and apparel items. (Chapter 12)

Oxygen Bleach. Color-safe bleach. (Chapter 12)

Packaging Sheet. Part of a technical package that details packaging specifications for the production and shipment of a garment. (Chapter 5)

Packaging Specifications. Detailed instructions for how garments should be packaged for shipping. (Chapter 5)

Pajamas or Pyjamas. One or more garments intended for sleepwear or nightwear and can be designed with or without feet. (Chapter 10)

Parallax Error. When the pointer on a dial is read from a slight angle rather than directly in front. (Chapter 11)

Pattern. A repetitive design that is created by weaving, knitting, felting, dyeing, or printing to add aesthetic interest to apparel products. (Chapter 2)

Perceived Quality. The consumer's opinion of the level of superiority of a product based on brand reputation, value, and the ability of the product to meet the expectations of the wearer. (Chapter 1)

Perchloroethylene/Perc/Tetrachloroethylene. Drycleaning solvent composed of halogenated hydrocarbons. (Chapter 12)

Performance. The ability of a product to function as it is intended. (Chapters 1 and 5)

Performance Features. The functional characteristics related to the intended use of a garment. (Chapter 1)

Performance Specification. Standards used to determine requirements for acceptability of materials or sewn products based on results obtained through testing. (Chapter 12)

Petrochemicals. Substance derived from petroleum used as primary components of synthetic detergents for laundering. (Chapter 12)

Petroleum. Drycleaning solvent composed of hydrocarbon distillates. (Chapter 12)

Physical Attributes. The design, material selection, construction, and finishing of a product. (Chapter 1)

Pictogram. Illustrations of body forms marked to indicate where specific body measurements (POM) should be taken. (Chapter 5)

Pigments. Add color to the surface of fibers and materials with the assistance of binding agents. (Chapter 3)

Pile Weave. Fabric woven with an additional set of yarns (warp or filling) in the base and loops on the surface of the fabric that can be cut or remain intact. (Chapter 3)

Pin Holes. Punctures along the selvedge of the fabric. (Chapter 13)

Pin Tucks. A single narrow fold or series of repeating parallel folds that are evenly spaced and stitched down from seam line to seam line or within a garment section. (Chapter 4)

Piped Tucks. Tucks that have cording inserted into the tuck. (Chapter 4)

Placket. A finished opening within a garment area that allows for the wearer to easily put the garment on or remove it. (Chapter 4)

Plain Surface Materials. Smooth surface fabrics that do not have a raised surface from fibers or yarns or fancy knit or woven structures. (Chapter 10)

Plain Weave. Woven fabric created using two harnesses with a shuttle passing the filling yarns over and under the warp yarns, alternating with each row. (Chapter 3)

Pleats. Fabric folded back on itself along the grainline to provide a means for fitting contours of the body and for design interest. (Chapter 4)

Ply Adhesion or Ply Security. The ability of a thread to maintain its structure without unraveling during sewing. (Chapter 6)

Ply Yarns. Two or more yarns twisted, wrapped, entangled, or chemically bound together. (Chapter 3)

Points of Measurement (POM). The measurement points indicated on a technical flat sketch of a garment. (Chapter 5)

Polarizer. Part of a microscope that allows the maximum amount of light to be transmitted through a specimen. (Chapter 11)

Precision. The agreement of data obtained from observations collected during testing. (Chapter 11)

Precision and Bias. Statistical data used for determining if a significant difference exists and if other factors may influence test results. (Chapter 11)

Price. The amount designated by the seller to be paid by the consumer in exchange for a product or service. (Chapter 1)

Price Point Classification. The range of prices, lowest to highest, upon which competitive products are offered in the marketplace. (Chapter 1)

Primary Dimension. The most important body or product dimension used to define the size of the item. (Chapter 5)

Primary Quality Indicators. The intrinsic attributes that comprise the physical structure of a product. (Chapter 1)

Princess Line Seam. A vertical seam that intersects the bust apex and extends to the shoulder or armhole. (Chapter 4)

Procedure. Instructions for how to perform a standard test. (Chapter 11)

Process Layout or Skill Center. A factory layout with equipment organized into pods of work areas. (Chapter 8)

Product Benefits. The physical attributes and performance features consumers desire to meet their needs and expectations. (Chapter 1)

Product Data Management. Software that is integrated with PLM to allow digital files for apparel products to be shared along the supply chain. (Chapter 8)

Product-Defined Quality. The inherent measurable physical features and attributes of an apparel item that determined its level of excellence. (Chapter 1)

Product Layout or Line Layout. Factory layout of equipment arranged in assembly lines. (Chapter 8)

Product Lead Sheet/Design Sheet. The cover sheet for a technical package that provides preliminary information about the garment style. (Chapter 5)

Product Lifecycle Management (PLM). Software systems for managing and communicating information along the supply chain from start to finish. (Chapter 8)

Product Market Research. Gathering and analyzing information regarding competing products, innovative products being developed or introduced, and trend forecasts to provide insight for ways to improve existing products or to provide new opportunities in the marketplace for expanding business. (Chapter 2)

Product Recall. The action of removing defective products from the distribution chain and retrieving them from consumers to avoid harm to the customer due to a serious safety hazard. (Chapter 13)

Product Specifications. Standards for intrinsic components of a completed product such as size and fit, garment assembly, finishing, labeling, packaging, quality, and performance. (Chapters 1 and 5)

Product Value. The customer's perception of quality in relation to the price paid. (Chapter 1)

Production Pattern. The final pattern for a style that has been tested and perfected; includes seam and hem allowances, notches, grainlines, perforations, and pattern identifications. (Chapter 2)

Production Samples. Final sewn garment prototypes constructed with the exact same materials and methods that will be used for mass production of the garment. (Chapter 5)

Production System. Resources and sequencing of workflow in a factory needed to assemble a garment from beginning to end. (Chapter 8)

Progressive Bundle System. A production process that utilizes assembly line methods and garment bundles that are moved manually according to the sequential order of machines used to complete the garment. (Chapter 8)

Proportion. The harmonious interrelationships of the position and scale between silhouette, design, details, style lines, and fabric grain within a given design. (Chapter 2)

Protease Enzymes. Protein molecules that help remove protein-based stains. (Chapter 12)

Prototype. A sample of a garment design developed in a physical form or virtual environment to provide a means for testing fit, function, and overall aesthetic appeal. (Chapter 5)

Psychrometer. Device that measures the temperature and relative humidity of the air. (Chapter 11)

Quality. An individual's perception of a garment's inherent characteristics and performance; the level of excellence or superiority of a product in relation to others in the marketplace; the ability of a product to function for its intended use and be free of defects. (Chapter 1)

Quality Assurance. The method for managing and controlling the processes for development and manufacturing of apparel to ensure product quality and compliance with safety regulations. (Chapter 1)

Quality Control. The process for ensuring specified standards for quality are maintained through continual testing at different phases of production, performing frequent inspections, and ensuring proper use of equipment and established procedures. (Chapter 1)

Quick Response Manufacturing (QRM). A strategy for manufacturing and delivering goods to the consumer faster by eliminating or reducing any handling that does not add value in an effort to improve product quality, reduce costs, and provide the brand with a competitive edge. (Chapter 8)

Raised Fiber Surface Materials. Fabrics made with a raised surface from fibers or yarns or fancy knit or woven structures having a pile, nap, tuft, or flocked surface. (Chapter 10)

RAPEX. The European Union's raid alert system for consumer product safety. (Chapter 10)

Raw Edges. Unfinished frayed ends of fabric at hemlines or seams in a garment where the material is not stitched properly. (Chapter 13)

Raw Materials. Fibers, yarns, and unfinished fabrics. (Chapter 3)

Raw Materials Inspection. The act of assessing fabric to document flaws and their specific locations within a roll to determine if the quality level meets required standards. (Chapter 13)

REACH. European Community Regulation that stands for Registration, Evaluation, Authorization, and restriction of CHemical substances that regulates the safe use of chemicals and publishes a directive for restricted substances. (Chapter 10)

Reagent. Substance or compound used for producing a chemical reaction. (Chapter 11)

Receiving Inspections. The act of evaluating shipments of piece goods and component parts through testing, visual evaluation, and measurement of raw materials. (Chapter 13)

Reduced Inspection (Level I). A lowered level of inspection that can be instated when ten consecutive shipments or lots are meeting acceptance quality limits under level II normal inspection. (Chapter 13)

Referenced Documents or Normative References. Additional standards that are related to the selected test method. (Chapter 11)

Refractive Index. Speed at which polarized light travels through a specimen when compared to the speed of light in a vacuum. (Chapter 11)

Refurbishment. The process of cleaning and restoring the appearance of a garment. (Chapter 1)

Registration Number (RN). A number issued and registered by the U.S. Federal Trade Commission for the purpose of identifying U.S. manufacturers, importers, or distributors of apparel products. (Chapter 9)

Relative Humidity (RH). The amount of water vapor present in the air that is represented as a percentage of the maximum volume that the air can support at any specified temperature. (Chapter 11)

Release Tucks. A single fold or series of repeating parallel folds that are evenly spaced and stitched down for a designated length and then released to direct fullness to a particular area of a garment. (Chapter 4)

Reliability. Consistent, dependable, and reproducible test procedures and results obtained during testing. (Chapter 11)

Report. Documentation of procedures used for testing and calculation of data or standards used for rating, accompanied by a statement of results. (Chapter 11)

Reproducible. When testing is performed on the same material or garment, and the data collected are consistent. (Chapter 11)

Retail Price. The amount designated by the retailer to be paid by the consumer in exchange for a product or service. (Chapter 1)

Rhythm. The movement of the viewer's eyes through each part of a garment that is carefully planned. (Chapter 2)

Right-Hand Twill. Woven twill fabric having diagonal lines that extend from the lower left to the upper right. (Chapter 3)

Rolled Hem. A narrow hem finish where the raw edge of the material is fed through an attachment where it is folded and stitched. (Chapter 4)

Ropy Hem. Skewed or twisted fabric within the hem of a garment. (Chapter 13)

Ruching. A controlled predetermined amount of fullness that is gathered and released to correspond to parallel a seam in a repeating pattern. (Chapter 4)

Rule of Thirds. Dividing a design into three unequal portions in an effort to control where the viewer's eyes are drawn. (Chapter 2)

Run-off or Overrun Stitching. Top or edge stitching that was continued beyond the point where it was supposed to end. (Chapter 13)

Safety Precautions or Hazards. Precautions to be taken to ensure the lab technician's safety when conducting test procedures. (Chapter 11)

Sample Plan. Arrangement and procedure for obtaining test specimens from a sample. (Chapter 11)

Sample Size. A size designated by a manufacturer that typically falls in the middle of their size range offerings, such as a size 8 or 10 for misses. (Chapter 5)

Samples. Materials or garments selected for testing. (Chapter 11)

Sampling or Test Specimens. Section of a test method containing detailed procedures for taking and preparing specimens such as dimensions, the number required, and the methods for preparing them prior to testing. (Chapter 11)

Satin Weave. A woven fabric created with 5 to 12 harnesses having four or more yarns passing over before passing under one yarn. (Chapter 3)

Saturation. The sharpness or dullness of a color's intensity. (Chapters 2 and 4)

Scalloped Tucks. Tucks that are formed by drawing in the folded edge at evenly spaced intervals and catching it at the stitch line to create a shell-shaped edge. (Chapter 4)

SCC (Standards Council of Canada). The organization that administers the National Standards System in Canada. (Chapter 1)

Scope. Summary of the purpose of the test and the materials covered by the standard. (Chapter 11)

Scrimp. An undyed portion of a fabric print caused by the formation of a crease during printing. (Chapter 13)

Seam. Two or more layers of material joined together by means of stitching, ultrasonic sealing, laser welding, or thermal bonding. (Chapters 4 and 7)

Seam Allowance. Measured distance between the stitchline of the seam and the raw edge of the fabric of a garment panel. (Chapter 7)

Seam Class 1. ISO designation for assembling two or more plies of fabric of which two are limited in width on the side where the seam is constructed. (Chapter 7)

Seam Class 2. ISO designation for assembling two or more plies of material that are limited in the width on opposite sides from each other, where they join into the seam. One of the plies of fabric is positioned below the other; then they are overlapped. (Chapter 7)

Seam Class 3. ISO designation for assembling seams constructed with a minimum of one ply of fabric that is limited in width on one side and a binding that is limited in width on two sides and that wraps over the raw edge of the other ply of material to finish the edge. (Chapter 7)

Seam Class 4. ISO designation for seams that are constructed with a minimum of two pieces of material limited in width having the raw edges abutted and stitched to cover over the joint. (Chapter 7)

Seam Class 5. ISO designation for assembling seams constructed with a minimum of one ply of fabric that is unlimited on two sides and any other material integrated into the seam can be limited on one or two sides. (Chapter 7)

Seam Class 6. ISO designation for assembling seams constructed with one ply of fabric that is limited in width on either the right or left side. (Chapter 7)

Seam Class 7. ISO designation for assembling seams constructed with at least two pieces of material where one piece is limited in width on one side and the other piece is limited on two of the sides. (Chapter 7)

Seam Class 8. ISO designation for assembling seams constructed with a minimum of one ply of fabric that is limited in width on two sides and any additional components that are equally limited in width on two sides. (Chapter 7)

Seam Efficiency. The relationship between the strength of a fabric and the strength of the seam construction. (Chapter 7)

Seam Failure. Rupture of the stitching and thread or separation of a sealed, welded, or thermal bond where fabrics are joined in a garment. (Chapter 7)

Seam Grin. The result of stitch balance being too loose which causes the seam to gap when stress is applied; unwanted visible stitching within a seam on the exterior of the garment that is caused by unbalanced thread tension. (Chapters 7 and 13)

Seam Pucker. Wrinkling of fabric in the stitching line. (Chapter 13)

Seam Repair. Restitching to correct broken stitches that are part of a seam. (Chapter 13)

Seam Slippage. Tension applied to a garment seam causing the yarns of the material to slide away from the seam; the displacement of warp yarns within a material, which causes them to slide and eventually pull out of the seam. (Chapters 7 and 12)

Seam Strength. Amount of force that can be applied to a seam to rupture the stitching or bonding. (Chapter 12)

Seamless Garments. Three-dimensional preshaped garments that utilize knit-and-wear construction and do not contain seams. (Chapter 7)Secondary Dimension. An ancillary body mass dimension that is used in conjunction with the primary dimension to distinguish the size of the item. (Chapter 5)

Secondary Quality Indicators. Extrinsic attributes that are not part of the physical makeup of a product that influence a consumer's perception of quality. (Chapter 1)

Separate Facing. A facing that finishes the edges of one garment area (i.e., a neckline or armhole). (Chapter 4)

Serviceability. The ease of care for a garment; ability of an apparel item to retain its shape and appearance. (Chapter 1)

Sewing Thread. A thin strand or ply of flexible yarn made from staple fibers, single monofilament, multiple filaments, or cords that are bonded together. (Chapter 6)

Sew-through Button. A button having two or four holes that are used for attaching them to a garment. (Chapter 4)

Shaded. The difference in color across the fabric width. (Chapter 13)

Shading. Mismatched dye lots of garment components and subassemblies that affect the appearance of the finished product. (Chapter 13)

Shank Button. A button containing a protruding loop or hole that is used for stitching it to a garment. The point of attachment is concealed from the face of the garment. (Chapter 4)

Shaped Facing. A separate piece of material ranging from 1? to 2 inches (3.8 cm to 5.08 cm) in width that is stitched and folded back to hide the raw edges. (Chapter 4)

Shaping Devices. Structured pads, wires, boning, tapes, stays, and elastic used to provide architectural support for garments. (Chapter 3)

Shaping Methods (Shaping Strategies). Darts and dart equivalents used to fit a garment around the contours of the body. (Chapter 4)

Shrinkage. A decrease in the dimension of a garment after cleaning. (Chapter 12)

Side Panel. A vertical panel that extends to the armhole used in place of a side seam to provide more fit. (Chapter 4)

Significance and Use. Information regarding acceptance testing procedures and how to describe the data gathered. (Chapter 11)

Significant Figures. The number of decimal places needed when reporting test data to provide accuracy. (Chapter 11)

Silhouette. The overall shape or outline of a garment. (Chapter 2)

Singles Yarn. One continuous filament or staple fibers twisted together to form a spun yarn. (Chapter 3)

Size Defects. Problems with the gradation of sizes or differences in the measurement of one garment part to another. (Chapter 13)

Size Designations. Body size indicator for apparel items identified on a garment label as a letter code or number. (Chapter 5)

Size Specifications. Detailed information regarding specific dimensions and tolerances for a garment that are taken at designated measurement points based on the garment styling, fit, manufacturer size requirements, grade rules, size range, and production needs. (Chapter 5)

Sizing System. Body measurements classified by gender, age, and sometimes garment type to provide size classifications for mass-produced apparel. (Chapter 5)

Skew. Displaced yarns within a knitted or woven fabric. (Chapter 13)

Skipped Stitch. When a stitch is not formed along the seam line. (Chapter 13)

Slot Buttonhole. A finished opening in a seam that allows a button to pass through and remain secured. (Chapter 4)

Slub. A yarn defect where one section is visibly thicker than the rest. (Chapter 13)

Smart Cards or Electronic Bundle Tickets. A card with a magnetic strip that can be swiped and read by a card reader and that essentially contains the same information as bundle tickets and serves the same purpose as the bundle coupon to monitor work completed. (Chapter 8)

Snap. A fastener with a ball and socket that join together to close. (Chapter 4)

Soaps. Water-soluble cleaning agents made from fats and oils or fatty acids from animal or plant sources that are treated with strong alkali such as sodium or potassium. (Chapter 12)

Sodium C10–16 Alkylbenzenesulfonate. A type of surfactant. (Chapter 12)

Sodium Carbonate/Soda Ash. Alkalinity-builder in chlorine bleach. (Chapter 12)

Sodium Carbonate Peroxide. Primary active ingredient in powder oxygen bleach. (Chapter 12)

Sodium Chloride. Salt that is added to thicken and stabilize liquid bleach. (Chapter 12)

Sodium Hydroxide/Caustic Soda. Adjusts pH level of bleach and aids in removal of grease and alcohol based stains. (Chapter 12)

Sodium Hypochlorite. The primary active ingredient found in chlorine bleach. (Chapter 12)

Sodium Polyacrylate. Additive to chlorine bleach that prevents redeposition of soils. (Chapter 12)

Soil. Chalk, ink, dirt, oil, or grease that create a stain. (Chapter 13)

Soiled Yarn. A dirty yarn. (Chapter 13)

Solubility. Dissolving fibers in chemical substances or reagents to identify the fibers present in a material. (Chapter 11)

Sourcing. The location and assessment of resources to acquire materials or manufacturing needed for apparel products that meet the quality standards required; procuring resources to acquire materials or manufactured apparel products. (Chapter 8)

Sourcing Agent. The liaison between a manufacturer, factory, and retailer who helps oversee production and monitor quality based on product specifications. (Chapter 8)

Spaced Tucks. A single fold or series of repeating parallel folds that are evenly spaced and stitched down from seam line to seam line or within a garment section. (Chapter 4)

Spec Sheets. Part of a technical package that details size specification information for the production of a garment. (Chapter 5)

Specification Requirements. List of fabric and garment properties that should conform to the performance standard. (Chapter 12)

Specifications. Established requirements for determining whether a material or product satisfies quality standards related to performance criteria, safety, or physical, mechanical, or chemical properties. (Chapter 1)

Specimen. Fibers, yarns, sewing thread, material, garment sections, or component parts. (Chapter 11)

Spectral Data. The percentage of light reflected at each wavelength along the visual spectrum that is digitally measured using a colorimeter or spectrophotometer. (Chapter 2)

Speed-to-Market. The ability of a brand to quickly design, manufacture and distribute products by means of reducing lead times so products enter the marketplace sooner for consumer consumption. (Chapter 8)

Split Yoke. A horizontal panel with a seam down the center front or center back that is used to shape a garment at the seams where it is joined. (Chapter 4)

Spun Thread. Single yarns comprised of staple fibers that are twisted together. (Chapter 6)

SS Superimposed Seam Class. ASTM International designation for assembling two or more plies of fabric that are overlaid, one on top of another, with their raw edges aligned and stitched together at a specified distance from the raw edge, with one or more rows of stitching. (Chapter 7)

Standard Atmosphere for Preconditioning Textiles. 122°F + 2 degrees (50°C+ 1 degree) with relative humidity between 5 percent and 25 percent having a tolerance of +2 percent. (Chapter 11)

Standard Atmospheric Conditions. 70° Fahrenheit (+) 2 degrees (21° Celsius + 1 degree) and 65 percent relative humidity + 2 percent. (Chapter 11)

Standard Inspection Procedure (SIP). The guidelines for evaluating garments during inspection. (Chapter 13)

Standard Reference Laundry Detergent. A basic detergent formulation used in a laboratory setting for testing materials and garments. (Chapter 12)

Standardized Color Matching System. Systems used to specify and manage color throughout the design and manufacturing processes. (Chapter 2)

Standardized Test Methods. Procedures for conducting experiments, examining, and evaluating textile materials and apparel products. (Chapter 11)

Standards. Technical documents for test methods and specifications developed and established within the consensus of international, national, and federal organizations and agencies, consortiums, or individual companies as a method for producing repeatable results to increase product quality and safety. (Chapter 1)

Staple Fibers. Short fibers ranging in length from 3/8 inch (5 millimeters) to 19.5 inches (500 millimeters). (Chapter 3)

Statistical Process Control (SPC) or Statistical Quality Control (SQC). A system for measuring and controlling quality that focuses on improving processes and prevention of problems throughout all product-producing phases. (Chapter 13)

Stitch. The loop formation created by hand or machine to interlock one or more threads for the purpose of creating surface decoration or seams. (Chapter 6)

Stitch Length. The length of a stitch in measured in millimeters (mm). (Chapter 6)

Stitches Per Centimeter (SPC). The number of stitches measured within one centimeter. (Chapter 6)

Stitches Per Inch (SPI). The number of stitches measured within one inch. (Chapter 6)

Stitching. A series of repeating stitches. (Chapter 6)

Stitchless. Seams joined by means of thermal bonding, laser welding, and ultrasonic sealing rather than with thread and stitches. (Chapter 7)

Stoddard Solvent. Petroleum based drycleaning solvents. (Chapter 12)

Straight Buttonhole. A zigzag rectangular shaped opening that can accommodate a button to pass through and remain secured. (Chapter 4)

Straight Darts. The legs of the dart converge to a diminishing point creating line segments that are not curved. (Chapter 4)

Straight Grain. Runs in the warp or wale direction of a fabric and is parallel to the selvedge. (Chapter 4)

Straight Seam. Two or more plies of fabric sewn together from one point to another without any curves. (Chapter 4)

Straight Waistband. A straight band of material that is attached at the waistline of a garment to finish the top edge. (Chapter 4)

Stretch Waistband. Elastic is inserted into the waistband to provide elasticity. (Chapter 4)

Strike Back or Back-Bleed. The appearance of adhesive on the surface of the interlining inside the garment due to overliquefied resin during fusing. (Chapter 13)

Strike-off. Physical representations of printed fabric that are evaluated against color standards for accuracy of color matching. (Chapter 2)

Strike Through or Bleed-Through. The appearance of adhesive on the surface of the garment due to overliquefied resin during fusing. (Chapter 13)

S-Twist. Yarn turned or twisted in the right direction creating diagonal lines that extend from the upper left to the lower right duplicating the same line direction as the middle of the letter S. (Chapter 3)

Subassemblies. Fasteners, closures, stitches, and seams incorporated into a finished product. (Chapter 12)

Substrate. Fibers, yarns, and fabrics. (Chapter 2)

Summary of Test Method, Summary of Practice, or Principle. Portion of a test standard providing a concise explanation of the procedure to be performed. (Chapter 11)

Supply Chain. A network of suppliers, manufacturers, and retailers responsible for producing, handling, and distributing products. (Chapter 8)

Supply Chain Management. The integration of business functions combined with the expertise, resources, and organized efforts from suppliers, manufacturers, and retailers to manage the flow of raw materials through production, distribution, and sale to the end user. (Chapter 8)

Support Materials. Separate plies of fabrics used to reinforce portions of a garment to achieve desired silhouettes or shaping. (Chapter 3)

Surface Active Molecules/Surface Active Agents/ Surfactants. Molecules found in soaps and detergents that aid in wetting and cleaning. They are composed of two parts—a carboxylate head that is hydrophilic (water-loving) and a hydrocarbon chain tail that is hydrophobic (water-repelling). (Chapter 12)

Surface Embellishments. Materials applied to the surface of a garment to add decoration such as applique, beads, embossing, embroidery, foiling, rhinestones, screen printing, sequins, and trapunto. (Chapter 3)

Surface Flash. When a flame quickly spreads across the surface of the fabric but ignition of the base materials does not occur. (Chapter 10)

Symmetric (Formal Balance). When a garment is vertically divided into two equal sides that appear identical. (Chapter 2)

Synthetic Detergents. Cleaning substances composed of petrochemicals or oleochemicals, alcohol, and sulfate used for laundering many apparel products. (Chapter 12)

Tack Buttons. A metal button containing a shank that attaches to the garment by means of a tack with a prong. (Chapter 4)

Tailored Placket. Two strips of unequal lengths of fabric are stitched to encase the raw edges of the opening. The narrower strip is hidden while the wider strip shows on the face of the garment when closed. (Chapter 4)

Target Market Research (Consumer Trend Research). Demographic and lifestyle data gathered and analyzed based on a brands customer; the examination of demographic and lifestyle data for both existing and potential customers within specific market segments. (Chapters 1 and 2)

Tearing Strength. Amount of force that can be applied to a material to continue a tear. (Chapter 12)

Technical Advisory Group (TAG). The organization that represents the United States (U.S.) in all ISO TC 38 Textile Committee actions. (Chapter 1)

Technical Packages/Production Package/Tech Pack. Detailed documents that include important information pertaining to the development, assembly, and packaging of a garment. (Chapter 5)

Technician Error or Observer Error. When a scale is misread. (Chapter 11)

Tensile Strength. Ability to resist tension before rupturing or breaking. (Chapter 12)

Terminology. Important terms used within the test standard. (Chapter 11)

Test Method Name. Title of the test standard that distinguishes the type of test covered within the standard. (Chapter 11)

Test Method Number. Numeric configuration that identifies the test standard. (Chapter 11)

Test Methods. Specific technical procedures for conducting and gathering test results for identification, measurement, and evaluation purposes. (Chapter 1)

Testing. Procedures for evaluating materials, component parts, and completed garments. (Chapter 11)

Tex (tex). Mass per unit length of spun yarns from silk or manufactured fibers measured in grams per 1,000 meters. (Chapter 3)

Textile Fiber Products Identification Act (TFPIA). U.S. labeling regulation 16 CFR 303 for disclosing fiber content, registered identification number or identity of manufacturer, or importer, distributor, or retailer, and country of origin on garment labels. (Chapter 9)

Textile Finishes. Mechanical or chemical treatments to fabrics to change their aesthetic or performance properties to produce a desired effect for the end product. (Chapter 3)

Textile Goods Quality Labeling Regulations. Japanese labeling regulation requiring garment labels to accurately disclose fiber content, country of origin, water repellency, manufacturer identification information, and care. (Chapter 9)

Textile Labelling and Advertising Regulations (TLAR). Canadian labeling regulation C.R.C., c.1551 for listing required accurate disclosure of fiber content and identification of the manufacturer, importer, or distributer on garment labels. (Chapter 9)

Textile Products (Labelling and Fibre Composition) Regulations. Regulation enacted for the United Kingdom (UK) and all member states as well as the European Union (EU) for accurate disclosure of fiber content on garment labels. (Chapter 9)

Texture. The surface contour or visual interest that influences the aesthetic appearance of textile products. (Chapter 2)

Textured Thread. Continuous filament fibers that have been heat set to create texture and add bulk. (Chapter 6)

Thermal Bonded Seams. Seam components are glued together using thermoplastic film adhesives. (Chapter 7)

Thin Place. Fabric defect in which the yarns are thinner and more loosely spaced in comparison to the rest of the fabric. (Chapter 13)

Thread Discoloration. Change in the thread color of garment stitching caused by the pickup of excess dye from the fabric during wet processing. (Chapter 13)

Thread Size. The linear density or mass per unit length of thread; thickness or diameter of the thread. (Chapter 6)

Thread Specifications. Detailed information describing thread size and composition for each sewing operation of a garment, performance requirements, and color standards for manufacturing. (Chapter 5)

Thread Tension. The relationship among sewing threads that comprise stitching. (Chapter 6)

Ticket Number. The amount of raw fiber contained in an unfinished sewing thread. (Chapter 6)

Ties. Opposing pairs of ribbon or strips of fabric used to secure a garment. (Chapter 4)

Tightened Inspection (Level III). A heightened level that requires more shipments to be inspected due to the number of defects that exceed the acceptance quality limit. (Chapter 13)

Tight-Fitting Sleepwear. Children's sleepwear that fits very closely to the body and that is designed and produced based on maximum garment measurements to reduce the risk of the garment catching fire. (Chapter 10)

Tow. Filament fiber that has been cut into staple lengths. (Chapter 3)

Toxicity. A safety hazard resulting from chemical substances used in some apparel products that come in contact with the wearer's skin and cause skin sensitivity; some substances are carcinogenic. (Chapter 10)

Transfer Machines. Drycleaning equipment composed of a machine where the clothes were cleaned and then moved to another machine to be dried. (Chapter 12)

Trend Forecasting Services. Companies that predict and provide seasonal color, materials, and style direction based on research of global mega trends, political, social, and cultural trends, as well as consumer behavior, lifestyle changes, and stimuli from various types of media outlets. (Chapter 2)

Trim. Decorative linear materials attached to the surface of the garment or inserted and sewn into a seam to enhance the design such as braid, cording, fringe, lace, passementerie, piping (cordedge), and ribbon. (Chapter 3)

Trim Sheet. Part of a technical package that details trim specification information for the production of a garment. (Chapter 5)

Trouser Fly Zipper Insertion. A facing is stitched to the zipper and extends beyond the center to hide the other side of the zipper that is stitched to a fabric underlay. (Chapter 4)

Trouser Waistband (Curtain Waistband). A commercially prepared waistband material containing a reinforced bias strip is stitched to the top edge of the waistband and allowed to hang free at the bottom edge. (Chapter 4)

True Bias. Grain that runs at a 45° angle from the intersection of straight and cross-grains. (Chapter 4)

Tucks. Fabric that is folded back on itself along the grainline, stitched down completely or partially, and then released. (Chapter 4)

Turns Per Centimeter (tpcm). The amount of twist applied to a yarn measured in centimeters. (Chapter 3)

Turns Per Inch (tpi). The amount of twist applied to a yarn measured in inches. (Chapter 3)

Turns Per Meter. The amount of twist applied to a yarn measured in meters. (Chapter 3)

Twill Weave. Woven structure formed using three or more harnesses whereas the shuttle carrying the filling yarn crosses over two or more warp yarns then passes under one or more yarns creating a diagonal line to the right or left in subsequent rows. (Chapter 3)

Twisted Garment. Side seams of a garment that wrap to the front or back and distort the appearance. (Chapter 13)

Ultrasonic Sealing. High-frequency sound waves that are used to create friction between thermoplastic fibers of a fabric and that produce enough heat to melt the material to fuse it together. (Chapter 7)

Ultrasonic Slitting. High-frequency sound waves used to cut and seal fabric edges of thermoplastic fiber fabrics. (Chapter 7)

Unbalanced Plain Weave. Plain weave structure having different warp yarns than filling yarns in size, type, and amount. (Chapter 3)

Unbalanced Stitch Tension. Stitching that is too tight causing puckers along the stitch line or seam, or too loose causing seam grin. (Chapter 13)

Unbalanced Thread Tension. When the thread tension for stitching is too tight or too loose, impacting the appearance, smoothness, and performance of stitching or seam construction. (Chapter 6)

Underfusing. Weak bond between the interlining and shell materials due to insufficient heat, time, or pressure during fusing. (Chapter 13)

Uneven Seams. The measured difference in the length of seams that should be identical. (Chapter 13)

Uneven Twill. Unbalanced twill weave having more yarns on the face (known as *warp-faced twill*) or more on the back (known as *weft-faced twill*). (Chapter 3)

Unit Production System. A production method that utilizes assembly line principles wherein individual garments are loaded into carriers that are electronically transported from operation to operation until the garment is completely assembled. (Chapter 8)

Unraveling Buttonhole. Buttonhole stitching that has not been secured at the end or that has been severed. (Chapter 13)

Unrelated Seam. An unintended seam in a garment caused by an area that is caught in the seam. (Chapter 13)

User-Defined Quality. The ability of a product to meet individual customer's needs and wants in relation to their personal preferences for garment attributes and value. (Chapter 1)

Value (Luminance). The tints, midtones, and shade variations of a color. (Chapter 2)

Vanity Sizing. A method for sizing garments that appeals to an individual's ego by labeling a garment as a smaller size than its measured dimensions. (Chapter 5)

Virtual Prototype. Digital simulation of a garment developed using specialized software programs to convert two-dimensional pattern pieces into three-dimensional garments. (Chapter 5)

Visible Stay Stitching. Temporary stitching used to stabilize the fabric to prevent stretching or distortion from occurring during sewing that has not been removed or concealed within the garment. (Chapter 13)

Voluntary Standards. Test methods and specifications that are not enforced by law but are utilized to maintain quality standards developed by retailers, manufacturers, importers, government agencies, and consumers through consensus between all parties. (Chapter 1)

Wales. The vertical rows of knit stitches or loops that run in the lengthwise direction in knitted fabrics. (Chapter 3)

Warp-Faced Satin. Woven fabric containing warp yarns that float on the surface before passing under one filling yarn. (Chapter 3)

Water-Soluble Soils. Substances capable of dissolving in water. Examples include sugar, salt, syrup, dust, clay, and perspiration. (Chapter 12)

Water Spots. Discolorations on fabric caused by improper drying or contaminated water during dyeing. (Chapter 13)

Wavy Seam. A seam that does not lay flat. (Chapter 13)

Wear-Testing or Wear-Service Conditions. Garment samples that are worn and evaluated to determine performance and appearance characteristic of the item in use. (Chapter 11)

Weft-Faced Satin. Woven fabric containing filling yarns that float on the surface before passing under one warp yarn. (Chapter 3)

Wet Prints. Liquid dyes mixed with thickening agents to form a paste that is printed onto fabric to form a pattern on the surface. (Chapter 3)

Wetcleaning/Organic Drycleaning. Cleaning process similar to laundering that can be used for cleaning fabrics typically intended for drycleaning. (Chapter 12)

Wetting. The process that allows material to be immersed in water or an aqueous solution and become moistened or wet. (Chapter 12)

Wool Products Labeling Act of 1939. U.S. labeling regulation 16 CFR 300 for labeling garments containing wool outlining disclosing of fiber content, registered identification number or identity of manufacturer, or importer, distributor or retailer, and country of origin. (Chapter 9)

Wool Products Labeling (WPL) Number. A number issued and registered by the U.S. Federal Trade Commission for the purpose of identifying the manufacturer, importer, or distributor of apparel products. The FTC no longer issues WPL numbers. (Chapter 9)

Woolen Run (wr or Nw_e). Linear density in length per unit mass of coarse woolen and woolen blended yarns measured in 1,600 yard lengths per pound. (Chapter 3)

Work Flow. The planned sequence for the movement of a garment through the production process. (Chapter 8)

Work in Process or Work in Progress. The amount of raw materials or number of garments that are not yet complete at any given time during manufacturing. (Chapter 8)

Working Pattern. A pattern that is in the process of being evaluated for design, fit, and modified for production. (Chapter 2)

Worsted Count (wc or Nw_w). Linear density in length per unit mass of fine worsted wool and worsted wool blends, and acrylic yarns measured in 560 yard lengths per pound. (Chapter 3)

Wrinkle Recovery. Elasticity to bounce back after being subjected to compression or deformation. (Chapter 12)

Wrinkle Resistant. Materials ability to oppose deformations resulting from folding or bending. (Chapter 12)

Wrong Grain. Garment panels are not cut on the designated grain. (Chapter 13)

Wrong Thread Color. Mismatched thread color. (Chapter 13)

Wrong Thread Type and Size. Improper thread construction. (Chapter 13)

Yarn Construction. Structure of the yarn including yarn number, twist direction, and number of turns per inch, per centimeter, or per meter. (Chapters 3 and 11)

Yarn Number. Method for measuring a yarn's fineness. (Chapters 3 and 11)

Yoke. A horizontal panel used to shape a garment at the seam where it is joined used in lieu of darts. (Chapter 4)

Zipper. A fastener with molded plastic or metal teeth or coils that is used to open and close a garment. (Chapter 4)

Z-Twist. Yarn turned or twisted in the left direction creating diagonal lines that extend from the upper right to the lower left duplicating the same line direction as the middle of the letter Z. (Chapter 3)